Incorporating Science, Economics, and Sociology in Developing Sanitary and Phytosanitary Standards in International Trade

PROCEEDINGS OF A CONFERENCE

BOARD ON AGRICULTURE AND NATURAL RESOURCES

NATIONAL RESEARCH COUNCIL

NATIONAL ACADEMY PRESS
Washington, D.C.

NATIONAL ACADEMY PRESS · 2101 Constitution Avenue, NW · Washington, D.C. 20418

NOTICE: The project that is the subject of this report was approved by the Governing Board of the National Research Council, whose members are drawn from the councils of the National Academy of Sciences, the National Academy of Engineering, and the Institute of Medicine. The members of the committee responsible for the report were chosen for their special competences and with regard for appropriate balance.

This study was supported by Contract/Grant No. 43-3AEK-6-80107 between the National Academy of Sciences and the U.S. Department of Agriculture Economic Research Service. Any opinions, findings, conclusions, or recommendations expressed in this publication are those of the author(s) and do not necessarily reflect the views of the organizations or agencies that provided support for the project.

International Standard Book Number 0-309-07090-2

Additional copies of this report are available from National Academy Press, 2101 Constitution Avenue, N.W., Lockbox 285, Washington, D.C. 20055; (800) 624-6242 or (202) 334-3313 (in the Washington metropolitan area); Internet, http://www.nap.edu

Printed in the United States of America

THE NATIONAL ACADEMIES

National Academy of Sciences
National Academy of Engineering
Institute of Medicine
National Research Council

The **National Academy of Sciences** is a private, nonprofit, self-perpetuating society of distinguished scholars engaged in scientific and engineering research, dedicated to the furtherance of science and technology and to their use for the general welfare. Upon the authority of the charter granted to it by the Congress in 1863, the Academy has a mandate that requires it to advise the federal government on scientific and technical matters. Dr. Bruce M. Alberts is president of the National Academy of Sciences.

The **National Academy of Engineering** was established in 1964, under the charter of the National Academy of Sciences, as a parallel organization of outstanding engineers. It is autonomous in its administration and in the selection of its members, sharing with the National Academy of Sciences the responsibility for advising the federal government. The National Academy of Engineering also sponsors engineering programs aimed at meeting national needs, encourages education and research, and recognizes the superior achievements of engineers. Dr. William A. Wulf is president of the National Academy of Engineering.

The **Institute of Medicine** was established in 1970 by the National Academy of Sciences to secure the services of eminent members of appropriate professions in the examination of policy matters pertaining to the health of the public. The Institute acts under the responsibility given to the National Academy of Sciences by its congressional charter to be an adviser to the federal government and, upon its own initiative, to identify issues of medical care, research, and education. Dr. Kenneth I. Shine is president of the Institute of Medicine.

The **National Research Council** was organized by the National Academy of Sciences in 1916 to associate the broad community of science and technology with the Academy's purposes of furthering knowledge and advising the federal government. Functioning in accordance with general policies determined by the Academy, the Council has become the principal operating agency of both the National Academy of Sciences and the National Academy of Engineering in providing services to the government, the public, and the scientific and engineering communities. The Council is administered jointly by both Academies and the Institute of Medicine. Dr. Bruce M. Alberts and Dr. William A. Wulf are chairman and vice chairman, respectively, of the National Research Council.

Preface

As the world economy has moved toward more open trade under the Uruguay Round of the General Agreement on Tariffs and Trade (GATT), there has been an increasing focus on managing potential conflicts between a country's right to take measures to protect its citizens, production systems, and environment (including plant and animal species) from risks and the effects of such protection on trade. In the area of sanitary and phytosanitary (SPS) regulations, the concern is that domestic regulations ostensibly designed as a means to protect plants, animals, or people may actually be used to protect domestic industries and interests. International standard-setting activities, the SPS Agreement of the Uruguay Round of GATT, and ongoing bilateral and multilateral negotiations are part of a process through which countries are attempting to manage conflicts between protective regulation in the SPS area and open trade.

Progressive trade liberalization has increased the importance of managing SPS issues (e.g., quarantine policies, product and process standards) between countries as they seek to protect human, animal, and plant life and health from biological and chemical risks, while simultaneously facilitating trade. The SPS Agreement, which went into effect in 1995, itself defines a set of principles for this management and provides a forum for settling disputes within the World Trade Organization framework. However, the operation of that set of principles will only be fully defined through experience under the agreement. Furthermore, the acceptable relationship between SPS measures and trade is the subject of ongoing negotiation between countries through the standard-setting activities of international organizations and multilateral and bilateral trade discussions. Thus, we are in a period of active institutional innovation that is resulting in a revised set of international relationships.

In 1998, the Economic Research Service of the U.S. Department of Agriculture asked the Board on Agriculture and Natural Resources of the National Research Council to organize a conference to address the roles of science, economics, and culture in agricultural trade policy. The conference was to focus on how scientific standards could be applied to international trade agreements in the post-Uruguay Round era but also take into account critical nonscientific factors surrounding SPS standards and related technical barriers to trade. Specifically, the conference was to focus on: (1) the critical roles and binding limitations of science in assessing SPS barriers to trade; (2) the critical roles and binding limitations of economics in assessing SPS barriers to trade; (3) the roles of values, other socioanthropological factors, and associated politics in determining SPS barriers to trade; and (4) an analytical framework for incorporating science, economics, values, and politics in SPS decision making.

The conference was held on January 25–27, 1999, at The National Academies' Beckman Center in Irvine, California. The participants focused on the roles of the biological and natural sciences, economics, sociology, politics, and culture, and approaches in understanding and evaluating differences in risk perception, assessment, and management across countries; the impact of SPS measures on plant, animal, and food safety; and the relationship between SPS measures and open trade.

This report presents a synopsis of the two-and-a half-day event. The overview, which was prepared by Julie Caswell, provides a summary of the broad range of issues identified by the speakers and participants of the conference. I would like to thank Julie for her outstanding contributions to this volume both in this summary and in her thoughtful evaluations of the papers. I would also like to thank Timothy Josling, Raymond A. Jussaume, Jr., Peter Kareiva, D. Warner North, and David Vogel who assured the effort would be a success through thoughtful insights in the conference design and significant contributions during the meeting.

The concepts presented in the overview are the result of many excellent ideas that grew out of formal presentations and group discussions during the conference. Chapters that follow reflect views and opinions of individual authors.

It is conference organizers' hope that the ideas contained in this document and summarized in the overview, enlighten and inform future approaches to ensuring that scientific, cultural, and economic considerations are reflected in SPS standards in international trade.

V. Kerry Smith
Center for Environmental and Resource Economic Policy
Department of Agriculture and Resource Economics
North Carolina State University

ACKNOWLEDGMENTS

Many individuals contributed to organizing the conference, and to conference discussions and its proceedings. The Board on Agriculture and Natural Resources would like to acknowledge and thank the following individuals for their valuable assistance and contributions:

JULIE CASWELL, University of Massachusetts, Amherst
TIMOTHY JOSLING, Stanford University, Palo Alto, California
RAYMOND A. JUSSAUME, JR., Washington State University, Pullman
PETER KAREIVA, University of Washington, Seattle
D. WARNER NORTH, NorthWorks, Inc., Belmont, California
V. KERRY SMITH, North Carolina State University, Raleigh
DAVID VOGEL, University of California, Berkeley

The Board on Agriculture and Natural Resources would like to recognize the efforts of Mary Jane Letaw, Project Officer, during the early stages of this project. We also wish to thank Karen L. Imhof, Project Assistant, for her tireless assistance throughout this project, Stephanie Padgham, Project Assistant, for her work in the early stages of report preparation, Elaine MacGarraugh for editing the manuscript, and Laura Boschini, Project Assistant, for her efforts in preparing the final report for publication. Special thanks are due to Lucyna K. Kurtyka, Project Officer, for her dedication and continuing efforts in seeing this project through to completion.

This report has been reviewed in draft form by individuals chosen for their diverse perspectives and technical expertise, in accordance with procedures approved by the National Research Council's Report Review Committee. The purpose of this independent review is to provide candid and critical comments that will assist the institution in making the published report as sound as possible and to ensure that the report meets institutional standards for objectivity, evidence, and responsiveness to the study charge. The review comments and draft manuscript remain confidential to protect the integrity of the deliberative process.

We wish to thank the following individuals for their participation in the review of this report: Lawrence Busch, Michigan State University, East Lansing; Ricardo Godoy, Brandeis University, Waltham, Massachusetts; Neal Hooker, Colorado State University, Fort Collins; D. Gale Johnson, University of Chicago, Illinois; G. Edward Schuh, Humphrey Institute of Public Affairs, University of Minnesota, Minneapolis; Daniel Simberloff, University of Tennessee, Knoxville; Mitchell Small, Carnegie Mellon University, Pittsburgh, Pennsylvania; and Daniel Sumner, University of California, Davis.

While the individuals listed above have provided constructive comments and suggestions, it must be emphasized that responsibility for the final content of this report rests entirely with the authors and the National Research Council.

Contents

OVERVIEW, *Julie Caswell* 1
Conference Organization, 2
Current Institutions for Managing SPS Issues in International Trade, 4
Discussion of Issues Related to SPS Management in International Trade, 8
Summary, 19

1. **Historical and Social Science Perspectives on the Role of Risk
    Assessment and Science in Protecting the Domestic Economy:
    Some Background,** *G. Edward Schuh* 23
    Historical Perspective on Protectionism, 24
    A Perspective on Risk Assessment, 26
    The Importance of Adjustment Policies, 29
    Conclusions, 29
    References, 30

**PART I: AGRICULTURAL TRADE, RISK ASSESSMENT, AND THE
ROLE OF CULTURE IN RISK MANAGEMENT**

2. **Sanitary and Phytosanitary Risk Management in the Post-Uruguay
    Round Era: An Economic Perspective,** *Donna Roberts* 33

The SPS Agreement: Origin and Principal Provisions, 35
Cost–Benefit Analysis of SPS Regulations, 38
Is the SPS Agreement Congruent with Executive Branch Guidelines?, 43
Conclusions, 38
References, 49

3. **An Overview of Risk Assessment,** *John D. Stark* 51
Risk and Trade Barriers, 52
What is Risk Assessment?, 52
Selecting Toxicological Endpoints: What Do We Evaluate?, 55
Deterministic Risk Assessment, 56
Probabilistic Risk Assessment, 57
Protecting Humans, Plants, and Wildlife, 57
Risk Assessment of Genetically Engineered Organisms, 61
How Can We Be Fooled? Unprovable Risks, 61
Future Problems—Scientific Arguments About Risk Assessment, 62
Conclusions, 62
References, 63

4. **Technological Risk and Cultures of Rationality,** *Sheila Jasanoff* 65
Dimensions of Cross-National Variance, 68
Varieties of Cultural Explanation, 75
Conclusions, 81
References, 82

PART II: POLITICAL AND ECOLOGICAL ECONOMY

5. **Biological Impacts of Species Invasions: Implications for Policy
Makers,** *Karen Goodell, Ingrid M. Parker, and Gregory S. Gilbert* 87
Impact from an Anthropocentric Perspective, 88
Case Study 1: The Grape Root Louse Phylloxera—The Importance
    of Recognizing and Regulating Vectors, 89
Vectors, 92
Impact from an Ecological Perspective, 94
Case Study 2: The Mosquito Fish—When Anthropocentric and
    Ecological Perspectives Clash, 97
Predicting Outcomes of Species Introductions, 99
Case Study 3: The Crayfish Plague and The Signal Crayfish—Limits
    to Prediction When Species Interact Synergistically, 105
Setting Priorities for Management of Invasive Species, 108
Conclusions, 109
Acknowledgements, 110
References, 110

6. **Risk Management and the World Trading System: Regulating International Trade Distortions Caused by National Sanitary and Phytosanitary Policies,** *David G. Victor* 118
    Introduction, 119
    The SPS Agreement: Major Elements, 120
    International Standards, 128
    Other WTO Agreements: GATT 1994 and the TBT Agreement, 137
    The System at Work: Three Cases, 139
    Analysis of the System at Work, 151
    Summary, 165
    References, 168

7. **Accounting For Consumers' Preferences in International Trade Rules,** *Jean-Christophe Bureau and Stephan Marette* 170
    Sanitary and Technical Barriers, 170
    Technical and Cultural Differences and Domestic Regulations, 173
    Accounting for Consumer Concerns, 180
    What are the Solutions for Reconciling Consumer Concerns and
        International Trade Rules?, 182
    Economic Analysis and the Settlement of Disputes, 191
    Conclusions, 193
    References, 194

PART III: CASE STUDIES

8. **Case Study 1: Meat Slaughtering and Processing Practices** 201
    The Danish Approach to Food Safety Issues Related to
    Pork Products, *Bent Nielsen*, 201
        Danish Consumers' Perspectives on Food Safety, 201
        Consumer Requirements of Danish Meat, 202
        Welfare, 204
        Conclusions, 205
        An Update on the Danish *Salmonella* Reduction Program, 205
        References, 209
    International Harmonization under the SPS Agreement,
    Bruce A. Silverglade, 210
        References, 216

9. **Case Study 2: Plant Quarantines and Hass Avocados** 217
    Role of Science in Solving Pest Quarantine Problems: Hass
    Avocado Case Study, *Walther Enkerlin Hoeflich*, 217
        Methodology, 218
        Results and Discussion, 221
        Conclusions, 226
        General Considerations, 226
        References, 226

The Hass Avocado Case: A Political Science Perspective,
  David Vogel, 228
  Reference, 230

10. Case Study 3: Genetically Modified Organisms................................. 231
  An Overview of Risk Assessment Procedures Applied To Genetically
  Engineered Crops, *Peter Kareiva and Michelle Marvier*, 231
    Containment, 232
    The Principle of Familiarity, 233
    Small-Scale Risk Assessment Experiments, 234
    Monitoring and a Precautionary Approach, 235
    References, 236
  Approaches to Risk and Risk Assessment, *Paul Thompson*, 238
    References, 244

APPENDIXES

Appendix A   SPS Agreement.. 249
Appendix B   Conference Program 263
Appendix C   Program Participants... 269
Appendix D   Conference Participants 273

TABLES, FIGURES, AND BOXES

Tables

Table 3-1   Acute LD_{50} Values of Selected Common Chemicals 54

Table 3-2   Pesticide Residues in Agricultural Commodities, 1997 58

Table 5-1   Counts of Plant Species Native to the United States That
Have a Known Economic Importance.................................. 92

Table 6-1   Acceptances of the *Codex Alimentarius* Commodity
Standards 134

Table 8-1   Monitoring Results of the *Salmonella* Reduction Program........ 208

Table 8-2   Special Slaughter Fees per Finisher 209

Table 9-1   Statistical Analysis of Hass Avocado Fruit Fly Infestation
Levels in Relation to Percentage of Dry Matter.................. 222

Table 9-2   Susceptibility of Hass Avocado to *A. ludens, A. serpentina,*
and *A. striata*—Forced Infestations in Fruits Attached to
the Tree and in Fruits at Different Time Intervals After
Harvest (Uruapan, Michoacan, 1993–1994) 224

Figures

Figure 2-1   A Partial Equilibrium Model of the Welfare Effects of
Alternative Import Protocols... 40

Figure 3-1   Dose-Response Relationship... 54

Figure 5-1   Frequency Histogram of the log Response Ratio (lnRR) for
Ant Impacts 96

Figure 5-2   Worldwide Distribution of *Gambusia affinis* 98

Figure 5-3   Mean Effect Sizes of Insect Invaders on Resident Confamilial
Species Versus More Distantly Related Species 104

Figure 6-1   Elaboration of Food Safety Standards and Other Guidelines
by the *Codex Alimentarius* Commission and its
Subsidiary Bodies.. 130

Figure 9-1   Case Study: Hass Avocado Background Chart 220

Figure 9-2   *A. ludens, A. serpentina,* and *A. striata*—
Forced Laboratory Infestation of Hass Avocados
(Uruapan, Michoacan, 1993–1994)................................... 222

Figure 9-3   Seasonal Fluctuation of *Anastrepha ludens* Populations and
Minimum and Maximum Temperatures in the Hass
Avocado Production Region of Michoacan, 1993–1994..... 225

Figure 9-4   Seasonal Fluctuations of *Anastrepha ludens* Populations and
Hass Avocado Harvest Period in Michoacan, 1993–1994 225

Figure 9-5   Role of Science in Solving Quarantine Pest Problems............. 227

Box

Box 7-1   Methods for Estimating the Benefits of Sanitary and
            Technical Regulations 190

.

Overview

JULIE A. CASWELL
Department of Resource Economics, University of Massachusetts

The rapid expansion of international trade has brought to the fore issues of conflicting national regulations in the area of plant, animal, and human health. These problems include the concern that regulations designed to protect health can also be used for protection of domestic producers against international competition. At a time when progressive tariff reform has opened up markets and facilitated trade, in part responding to consumer demands for access to a wide choice of products and services at reasonable prices, closer scrutiny of regulatory measures has become increasingly important. At the same time, there are clear differences among countries and cultures as to the types of risk citizens are willing to accept. The activities of this conference were based on the premise that risk analyses (i.e., risk assessment, management, and communication) are not exclusively the domain of the biological and natural sciences; the social sciences play a prominent role in describing how people in different contexts perceive and respond to risks. Any effort to manage sanitary and phytosanitary (SPS) issues in international trade must integrate all the sciences to develop practices for risk assessment, management, and communication that recognize international diversity in culture, experience, and institutions.

Uniform international standards can help, but no such norms are likely to be acceptable to all countries. Political and administrative structures also differ, causing differences in approaches and outcomes even when basic aims are compatible. Clearly there is considerable room for confusion and mistrust. The issue is how to balance the individual regulatory needs and approaches of countries with the goal of promoting freer trade. This issue arises not only for SPS standards but also in regard to regulations that affect other areas such as environmental quality, working conditions, and the exercise of intellectual property rights.

This conference focused on these issues in the specific area of SPS measures. This area includes provisions to protect plant and animal health and life and, more generally, the environment, and regulations that protect humans from foodborne risks. The Society for Risk Analysis defines a risk as the potential for realization of unwanted, adverse consequences to human life, health, property, or the environment; estimation of risk is usually based on the expected value of the conditional probability of the event occurring times the consequence of the event given that it has occurred.

SPS regulations that come under the purview of the World Trade Organization (WTO) SPS Agreement are those that (1) protect animal or plant life or health within a territory from risks arising from the entry, establishment or spread of pests, diseases, disease-carrying organisms, or disease-causing organisms; (2) protect human or animal life or health within a territory from risks arising from additives, contaminants, toxins, or disease-causing organisms in foods, beverages, or feedstuffs; (3) protect human life or health within a territory from risks arising from diseases carried by animals, plants, or products thereof, or from entry, establishment, or spread of pests; or (4) prevent or limit other damage within a territory from the entry, establishment, or spread of pests (see Appendix A for WTO SPS Agreement 1994, Annex A).

The task of this conference and of this report was to elucidate the place of science, culture, politics, and economics in the design and implementation of SPS measures and in their international management. The goal was to explore the critical roles and the limitations of the biological and natural sciences and the social sciences, such as economics, sociology, anthropology, philosophy, and political science in the management of SPS issues and in judging whether particular SPS measures create unacceptable barriers to international trade. The conference's objective also was to consider the elements that would compose a multidisciplinary analytical framework for SPS decision making and needs for future research.

CONFERENCE ORGANIZATION

A two-and-a-half-day conference was held in Irvine, California, January 25–27, 1999, to examine the roles of the biological and natural sciences, economics, sociology, politics, and culture in the management of trade issues related to SPS standards in the post-Uruguay Round era. Speakers and participants were drawn from across several disciplinary backgrounds: biology and natural sciences, sociology, economics, political science, and philosophy. They represented government agencies, universities, consumer and environmental groups, and producer organizations. Geographically, they represented the United States, Mexico, and the European Union. The conference program and questions for the breakout group discussions are presented in Appendix B. Program participants are presented in Appendix C, and conference participants are listed in Appendix D.

The main themes discussed at the conference were defined through a background discussion (Chapter 1), the topics selected for six commissioned papers (Chapters 2–7), and three case studies (Chapters 8–10). During his after-dinner presentation, G. Edward Schuh laid out historical and social science perspectives on protectionism, a perspective on risk assessment, and an emphasis on the importance of adjustment policies in the process of trade liberalization (Chapter 1). This presentation provided an intellectual context for discussions presented in the commissioned papers.

Each commissioned paper discussed one or more dimensions of how current institutions for SPS management perform and explored challenges for future management of the SPS process. The first group of commissioned papers (Chapters 2–4) addressed how the current system is operating and the challenges in managing SPS issues in international trade. The presenters were an economist, a natural scientist, and a social scientist. They discussed the current institutions for SPS management from an economic perspective (Chapter 2), the scientific issues faced in conducting risk assessments (Chapter 3), and cultural and political approaches to risk and its management (Chapter 4).

The second group of commissioned papers (Chapters 5–7) went into further depth on the biological, political, and economic questions that arise in SPS management. The presenters were a natural scientist, a political scientist, and an economist. They addressed the challenges in predicting the outcome and impacts of biological events (Chapter 5), the principles being developed by the world trading system to settle SPS-related disputes (Chapter 6), and how consumer concerns and the benefits and costs of regulation can be accounted for in SPS management (Chapter 7).

Case studies offer insights into the degree to which the different sciences and disciplinary approaches have been and are being integrated in the management of SPS issues in international trade. SPS cases have been gaining in prominence in recent years, particularly in the context of the dispute resolution process set up under the WTO.

SPS cases may be cross-classified by four major characteristics. The first is the type(s) of risk involved (e.g., to human, plant, animal, or environmental health). The second is by the trading partners involved in the case (e.g., developed–developed country trade or developed–developing country trade, trade within or between trading blocs). The third characteristic is the degree of current resolution of the case (e.g., settled versus on-going). Finally, cases are distinguished by the type of action trading partners have taken to manage the issues. These actions might include voluntary bilateral negotiation of equivalency; multilateral standard setting, for example through *Codex Alimentarius* Commission (*Codex*); or disputes, for example at the WTO.

Three cases that cut across the four major characteristics discussed above were selected as case studies for the conference. The first case study (Chapter 8) focused on the management of SPS issues related to international trade in meat products. Here countries are concerned with evaluating each other's process standards for production, slaughtering, and processing operations and final product standards. International trade in meat products may pose human or animal health risks. To date, most SPS management conflicts and efforts at regulatory rapprochement in this area have been among developed countries but they also occur in developing–developed country trade. This is an on-going SPS case area where the main approach to management has been cooperative through international standard setting and bilateral negotiation.

The second case study (Chapter 9) focused on plant and food product quarantines, with a specific discussion of changes in the U.S. quarantine policy for Hass avocados being imported from Mexico. Here the risk is phytosanitary and the trading partners involved are a developing and a developed country. The case was resolved through bilateral negotiation between the United States and Mexico in the context of the North American Free Trade Agreement (NAFTA).

The final case study (Chapter 10) discussed the international management of SPS issues related to the use of biotechnology, or genetically modified organisms (GMOs), in the agricultural, food, and fiber sector. The use of biotechnology may pose human, plant, animal, or environmental risks. Management of SPS issues related to biotechnology is a current focus of attention across the developed and developing countries. It is being negotiated between countries in bilateral and multilateral fora, and it is often speculated that GMO management could result in a WTO trade dispute between the United States and the European Union.

A natural scientist was paired with a social scientist in presenting each of the three case studies. The scientists were asked to address the following questions: How and how well are the sciences used in managing (or disputing) SPS differences between countries? How do cultural values, beliefs, and politics influence the ways that different countries and regions approach issues of risk analysis for SPS issues? How might the various sciences be better integrated or be used in a more complementary fashion in risk analyses used for SPS decision making? The speakers' perspectives on these issues in their case study areas are presented in Chapters 8–10.

CURRENT INSTITUTIONS FOR MANAGING SPS ISSUES IN INTERNATIONAL TRADE

Background

The key principles for the management of SPS issues in international trade were negotiated in the writing of the Agreement on the Application of Sanitary and Phytosanitary Measures (the SPS Agreement), which was negotiated in the Uruguay Round of the General Agreement on Tariffs and Trade (GATT) and is administered by the WTO. The SPS Agreement is intended to improve the climate for trade in agricultural and food products by specifying the mutual obligations of countries to avoid unnecessary trade impediments. The agreement rests on, and interacts with, a broad base of international standard-setting activities and developments in national, bilateral, and multilateral regulatory programs.

Early efforts at trade reform focused on reducing tariff barriers. In recent years, social regulation designed to protect people and the environment has been transformed dramatically in the United States and many other developed economies. Today, each developed country has an extensive system intended to protect its citizens and the environment, whereas countries in the developing world typically have less extensive systems. The result of these transformations in regulatory activities is a diverse array of policies across countries that address the same goals—protection of citizens and the environment. The SPS Agreement builds on a long history of international efforts to base domestic SPS regulatory programs on similar principles. However, the ongoing challenge is whether particular regulations that are acceptable and desirable to domestic stakeholders are likely to be compatible with an international regime designed to facilitate trade.

The need for an SPS Agreement had been felt for some time prior to its implementation in 1995. Trade conflicts in this area were becoming increasingly difficult to resolve. Existing trade rules gave countries the freedom to control imports (e.g., through use of bans, inspection systems, or labeling requirements) to protect human, animal, and plant health, but also to choose the basis for imposing such trade controls. As a result, a number of conflicts had arisen over the years, which were not easily resolved by the existing institutions.

One such conflict was the complaint by the United States and Canada about the European Union's rules, dating back to 1988, banning the importation of beef from cattle treated with growth-enhancing hormones. The SPS Agreement was aimed at just such disputes, which could not be effectively addressed under prior trade agreements because there was no binding process for the settlement of disputes. The prospect was for both a larger number and more extensive conflicts in the future. The beef hormone issue, although very prominent, involves a relatively small amount of trade. In contrast, the increasing use of genetically modified soybeans and corn in the United States implies that much of the supply of these products to the food processing industry will be from transgenic crops. If major overseas markets block the importation of foodstuffs incorporating the products of GMOs, then the resulting trade tensions will dramatically overshadow the beef hormone issue.

Differences in attitudes toward the use of hormones in beef production and toward the use of GMOs in foodstuffs are striking examples of a more general issue. Can cultural differences among countries for evaluating threats to people and the environment be managed within an international trading system? Do differences in the characterization of the events at risk and subjective perceptions of the extent of the risk and whether it is worth taking call into question a single approach to risk management? Is a primary emphasis on science appropriate as a way of reducing conflicts? Or does it merely transfer the conflict to other stages of the decision-making process? And how should social science, political economy, and other considerations be incorporated into decisions? Politics obviously play a role in both the domestic regulations and the trade tensions. How does one allow politics to translate the desires of consumers and producers through normal democratic processes without abandoning the trade system to control by special interests and distortion by those with other agendas? The conference participants sought to address these questions regarding the ways of regulating the global food industry to protect against the spread of disease, while at the same time maintaining the ability of countries to differ in tastes and cultural conventions.

Provisions of the SPS Agreement

The SPS Agreement is the strongest current international management tool for addressing SPS issues. It attempts to specify a framework within which individual countries should design their SPS measures (see Chapters 2 and 6). It is believed that if all countries adopt this framework, trade conflicts can be resolved in a more routine fashion. The framework embedded in the SPS Agreement has at its core the concept of risk analysis, although it does not fully adopt the three-part paradigm of risk analysis (risk assessment, management, and communication). The agreement focuses on the use of risk assessment as a necessary element in a country's choice of SPS measures that, consequently, have an impact on market access for imported products. Countries are free to choose their appropriate level of protection against imported pests and pathogens, but their regulations must be demonstrably based on an assessment of risk and clearly related to the control of the risk. Thus, the SPS Agreement seeks to harmonize analytical frameworks for addressing risk but not necessarily the level of protection required in each country. SPS decisions based on this model can then be regarded as "safe" from challenge by trading partners. The basic logic that underlies the agreement is that the use of science and risk assessment will provide an adequate basis for managing SPS risks. SPS measures that are maintained without such evidence can be challenged (SPS Agreement, Article 2.2).

Several other provisions of the SPS Agreement play a significant role in the design of domestic regulations in the plant, animal, and human health area. They require that SPS measures not discriminate in an arbitrary fashion among WTO members, or in a way that constitutes a disguised restriction on international trade (Article 2.3). The use of international standards is strongly endorsed (Article 3.1), notably those of the *Codex*, the International Office of Epizootics (OIE), and the regional and international organizations operating within the framework of the International Plant Protection Convention (IPPC). Using these standards relieves the country of the threat of challenge by other countries. Although countries can set standards that exceed the international norms, once again these must be based on scientific evidence (Article 3.3). In addition, countries are encouraged to accept as equivalent to their own the standards of exporting countries, which give the same level of protection, albeit by other means (Article 4.1). The concept of "pest- and disease-free" zones is recognized as a useful way of managing risks to plants, animals, or people while facilitating trade. These provisions rely on identifying areas that an exporting country can demonstrate to be free of a particular pest or disease (Article 6). Notification procedures are also established to increase the transparency of the regulations (Article 7).

From the point of view of clarifying the use of science and the place of economics, culture, and politics in SPS management, the key provisions of the agreement are contained in Article 5. This article attempts to define the requirements for the "assessment of risk" and the determination of the "appropriate level of protection." Under the agreement, risk assessment typically involves the identification of the hazard, appraisal of the likelihood of the consequences of the hazardous situation, and specification of the way in which the SPS measure reduces those consequences. This framework structures the issues as if the whole process falls within the realm of science, although it is, of course, acknowledged that scientists may disagree over aspects of any particular risk assessment and that there may be significant gaps in knowledge. However, the provision that a country should establish an "appropriate level of protection" allows other factors to be considered in defining comparable levels of safety. Specifically, in the case of plant and animal health (although not human health), the SPS Agreement allows the country to take into account economic and biological factors (Article 5.3). The country is left to decide the appropriate level of protection, thus giving the opportunity for the expression of political and cultural differences. All SPS measures must still be based on the assessment of risk even if they take into account other considerations, and measures should be used consistently so that they do not provide arbitrarily higher or lower levels of protection in different cases (Article 5.5). SPS measures should also not be more trade restrictive than necessary (Article 5.6). Where scientific data are not yet available to determine the appropriate policy, interim measures may be adopted until the required information is available (Article 5.7).

Other provisions of the WTO have implications for the management of SPS issues. Most important is the concept of national treatment, which requires a country to apply the same rules to domestic and imported products. In addition, the Agreement on Technical Barriers to Trade (TBT Agreement) sets criteria for technical regulations that affect trade. Although the TBT Agreement does not apply to SPS measures, it may come into play for a particular regulatory measure that does not have safety and health implications. For example, a measure such as a set of restrictions on the marketing of food products produced with the use of GMOs could be considered an SPS matter if the restrictions' goal is related to plant, animal, or human health, but a TBT issue if it is not.

The SPS Agreement has been the basis for several disputes over the course of its operation. The panels, which adjudicated the disputes, and the appellate reports that reviewed the panels' findings, have added valuable case law that helps in understanding the likely consequences of the agreement. The most important of these cases has been that concerning the European Union's ban on the import of beef treated with hormones (the hormone case). In its ruling, the WTO Appellate Body emphasized the central importance of the obligation to base public health measures in the food sector on an objective assessment of risks. The European Union was found (*inter alia*) not to have based its import ban on such a risk assessment. The panel also recognized that science cannot in itself dictate the SPS measures taken. But the measure had to have a "rational relationship" with the objective risk assessment.

As important as these dispute settlement results have been, probably more important have been the efforts that national governments have taken to review their SPS measures and alter them to establish greater conformity with the SPS Agreement. They have done so to avoid possible challenge, but also more broadly to demonstrate good faith compliance with the agreement and encourage others to comply. Such changes give hope that the elaboration of the rules for SPS measures in the agreement will contribute significantly to a reduction in trade conflicts and greater transparency in global food and agricultural markets.

At this point it is also useful to highlight what the SPS Agreement does not do or is not designed to do. The SPS Agreement does not formally use the full language of risk analysis currently employed in international regulatory circles. That language breaks the risk analysis process into three steps: risk assessment, risk management, and risk communication. As Donna Roberts noted in Chapter 2, the SPS Agreement does not use the term risk management, but instead refers to a country's choice of an "appropriate level of protection." The language describing risk assessment in the SPS Agreement presupposes integration of the elements of both conventional risk assessment and management as they have been used in the rule-making case law for domestic regulation of risks. Early cases have focused on the adequacy of the risk assessment providing the basis for an SPS measure. It is also important to acknowledge that the agreement requires that the SPS measure adopted be the least trade restrictive. Unfortunately the provisions of the agreement do not provide clear criteria for judging regulations based on this criterion. The agreement also includes some discussion of dimensions of an economic analysis that could underlie evaluation of SPS restrictions, but does not endorse or require the use of cost–benefit analysis as a component of a country's SPS decision making.

As part of WTO, the SPS Agreement has a primary focus on trade facilitation, while protecting the rights of countries to provide a level of protection that they deem to be appropriate. The agreement, disputes settled under it, and the ongoing and future trade negotiations all contribute to the development of precedents analogous to domestic rule making and judicial processes for regulation. These activities serve as fora for countries to define common principles for the management of SPS issues in international trade. In themselves, however, the SPS Agreement and the WTO process are not the primary international vehicles for such goals as improving SPS safety levels. The extent of protection is the venue of the international standard-setting bodies and bilateral and multilateral negotiations.

DISCUSSION OF ISSUES RELATED TO SPS MANAGEMENT IN INTERNATIONAL TRADE

The SPS arena is distinguished by the broad array of risks addressed within it. These risks range from control of the incidental importation of Asian long-horn beetles on wooden pallets and the levels of pathogens in packaged meat products to the possible plant, animal, and human risks of GMOs. The choice by governments of measures to address these risks and the international management of the trade consequences of those choices necessarily require analytical approaches from a range of disciplines. In the first years of experience under the SPS Agreement, risk assessment based on the biological and natural sciences has been the most prominent discipline in this process. However, cultural, political, and economic concerns have also shaped the experience. Some critics argue that the current approach may overemphasize scientific risk assessment at the expense of other important considerations. Risk assessment, particularly if too narrowly conceived, may be an inadequate basis for managing SPS issues and international relations. For example, a conference participant commented on an apparent contradiction in some current approaches to managing SPS issues, noting that as the social and life sciences are recognizing that they need to be more holistic and integrative, trade agreements can be subject to a reductionist perspective by putting complex, multidimensional issues in separate boxes.

The management of SPS issues in international trade requires an integrated framework. This involves broadening the risk assessment approach taken in the SPS Agreement to include explicitly the three elements of risk analysis: risk assessment, risk management, and risk communication. This may involve explicit consideration of differences in what constitutes an appropriate risk analysis for different types of SPS issues (e.g., threats to plant, animal, human, or environmental health). It may also include greater emphasis on comparative risk evaluations where the events and process associated with risks can be compared. And it may be coordinated with cost–benefit analysis to measure the full welfare implications of SPS measures and gauge their economic importance in relation to the net benefits from liberalizing international trade. A discussion of these points follows.

Broadening Cultural Perspectives for Systematic Risk Analysis

Culture was a major underlying theme in many of the presentations at the SPS conference. Social scientists have long recognized that an appreciation of the role of culture is crucial in any comprehensive risk analysis. As Sheila Jasanoff noted in her paper, "Divergent responses to risk, in particular, point to the ability of social norms and formations—in short, culture—to influence deeply the ways in which people come to grips with the uncertainties and dangers of the natural world" (Chapter 4).

Jasanoff described several ways in which these coping strategies can manifest themselves. First, framing is used to put boundaries on a problem that in principle can be solved. However, it is also important to acknowledge that, depending on the frame selected, different nations may vary in their response to the same threat. For example, she noted that Western scientists often attribute climate change to global emissions of greenhouse gases, whereas activists in developing nations tend to blame centuries of unsustainable practices by industrialized nations as the root cause of the problem. Second, Jasanoff argued that nations vary in style of regulation, which is often exhibited in processes used to provide interaction and communication between the bureaucracy and the population being impacted by policy. The United States, in particular, has a regulatory process that is more formal in terms of soliciting and processing input. At the same time, many disputes are more adversarial and require litigation for resolution. In other nations, affected parties may be expected to react directly to a perceived risk. Differences in the framing and style of regulation also accompany divergences in the types of evidence that governments and the public consider suitable for use in decision making. A striking contrast between nations is the emphasis on quantitative risk assessment in the United States to estimate the risks and uncertainties of cancer versus the use of more qualitative, weight-of-evidence approaches in Europe. With increased globalization and participation of disparate cultures in international trade, Jasanoff predicted that risk debates will become more numerous. Even among nations that appear to be closely similar in economic, social, and political aspirations, divergences in conceptualization and management of risk may preclude convergence in regulatory policies.

Despite the recognition by science of the importance of cultural differences in risk evaluation, the SPS Agreement provides no guidance and offers little scope for incorporating cultural analysis into SPS trade issues. The chapter by Jean-Christophe Bureau and Stephan Marette identified a series of questions about how culture affects risk assessment that are very difficult to resolve (see Chapter 7). They noted that it seems straightforward to acknowledge the cultural right of Islamic nations to erect trade barriers to pork imports. However, the U.S. beef exporters have been less willing to accept the fact that a large percentage of European consumers may have a "cultural aversion" to eating beef produced with growth-enhancing hormones or antibiotic drugs. Similarly, producers of French specialty cheeses made under traditional systems that have been codified under French law, and which utilize unpasteurized milk, take exception to U.S. policies that prohibit the importation of those cheeses on the grounds that they are not perceived to be safe.

A related theme that emerged from the conference, and one that was not addressed by the framers of the SPS and related agreements, is that culture also influences the "best scientific" risk assessments. Scientists are embedded within their own cultures, and their own cultural background influences their work. One example of this is whether "risk control" or "risk elimination" is selected as an appropriate scientific strategy for responding to a particular food safety problem. For example, in 1997 a number of people became seriously ill in the state of Washington as a result of exposure to *Salmonella typhimurium* DT-104. This exposure was traced to the eating of traditional-style *queso fresco* cheeses made from unpasteurized milk. The response chosen by consulted university scientists was to eliminate the risk by developing a new *queso fresco* recipe that could utilize pasteurized milk. This approach was successful, but an approach that was designed to control risk by developing a system similar to Hazard Analysis Critical Control Points for testing of nonpasteurized milk was not selected. One would guess that French food scientists would not suggest a program to develop new recipes for traditional French cheeses that would permit the use of pasteurized milk.

Culture can also influence the implementation of "best science" by affecting the way a scientist frames a research problem. This issue was discussed in John Stark's presentation on assessing risks to the environment associated with the introduction of exotic plants or insects to a particular ecosystem (see Chapter 3). He noted that there are many types of risk that can be evaluated as part of a risk assessment, such as hormesis (benefits of a small dose of a toxin), individual-versus population-level effects, effects on population structure, variation in life stage susceptibility, and the selection of specific measurement endpoints. There are also many ways that they can be evaluated. As a consequence, it is inevitable that professional judgment becomes part of the risk assessment process. The cultural superstructure in which a scientist lives undoubtedly influences his or her perception of what questions to ask and which risks to evaluate.

Incorporating culture into risk analyses adds a level of complexity to the process that many policymakers may prefer to avoid. However, citizen/consumer perceptions are an integral part of many SPS trade issues, and given that scientists must exercise personal judgments in conducting complex risk assessments, the issue of culture must be confronted. This suggests that risk assessments of important SPS trade issues will need to be both transdisciplinary and transnational if they are to be effective in contributing to the resolution of these issues. A participant at the conference posed the question, "For what purpose are we looking at cultural differences in SPS management? Are we looking to sustain them or to eliminate them?" Sheila Jasanoff responded that she was concerned that positions on risk analysis issues are justified as being purely rational, even though they contain unquestioned cultural elements. Cultural differences should persist at least long enough for people in different countries to understand analytical approaches to SPS management and evaluate whether they reflect their own underlying value commitments. At the same time, Jasanoff would not advocate a totally relativist position. There are analytical approaches that are better, but fostering those positions requires understanding of cultural differences and recognizes the importance of persuasion.

Science: The Challenges of Risk Assessment

The scientific and conceptual challenges involved in risk assessment for SPS issues were a second underlying theme of the conference. A particular emphasis was on these challenges in relation to ecological risk assessments. Risk analyses of all kinds typically struggle with data gaps, large uncertainties, the need to extrapolate, and the difficulty of quantifying risks associated with extremely unlikely events of large magnitude. Ecological risk analyses are, however, hampered by a unique set of additional exacerbating challenges. There are few "model systems" in ecology (such as laboratory mice or bacterial cultures) that can be used for extrapolation. There is no widely accepted quantitative theory (such as the exponential decay of radioactivity) for making even the simplest of calculations. Ecology is often viewed as a science of "special cases," in which the details of history and contingencies override any possibility for generality. The case studies and presentations on ecological risk assessment illustrated the complexity of the underlying processes that create these impressions.

The Mexican Hass avocado was initially excluded from the United States on the premise that its importation could also lead to the importation of insects that were thought to feed on avocados in Mexico. Once these products were allowed in, it was assumed that the pests would attack California avocados. To address these risks, feeding trials were performed with the Hass avocado and the pests—a specific species of fruit flies (no other avocado or fruit fly species would have sufficed as "model systems" because plant–insect associations are highly specialized). As Walther Enkerlin Hoeflich suggested, observations needed to be done exactly at the locations in which the Hass avocados were grown (see Chapter 9). Thus, a risk analysis had to be conducted with Hass avocados and no other avocado, and the analysis had to be conducted in the state of Michoacan that is the primary export source for these avocados. Models and extrapolations played no role in this risk assessment. Instead, the matter was resolved by direct experimentation using the avocado variety of interest and the pests that were hypothesized to be likely culprits.

Even with such an empirical approach, there remained some ambiguity. In particular, one of the fruit fly species that was of concern thrived on the Hass avocado if the fruit was detached from the tree and the flies confined inside cages. However, under natural field conditions this species of fruit fly was never found to have attacked avocados (favoring alternative hosts). Hence, the risk was assumed to be negligible because avocados exported from the relevant region of Mexico were never infested and, thus, were unlikely to be a source of infestation to California. This is an interesting case study, because if one relied solely on the caged experiments using detached avocados (which might be thought of as a "model system") then one would conclude that there was substantive risk associated with importing this crop. Only when the cage experiments were combined with detailed natural history observations in the field was the risk found to be minimal.

The discussion of the ecological impacts associated with biological invasions presented by Karen Goodell, Ingrid Parker, and Gregory Gilbert dealt with planned and unplanned introductions of species (see Chapter 5). Although only agricultural risks were considered when deciding whether to allow importation of Hass avocados, recent research has indicated the importance of considering a wide range of ecological risks that might arise when a non-native species is introduced into a country. The diverse array of examples presented by Goodell et al. reveal a familiar theme: Few model systems can be identified, each biological invasion seems to have a unique story, and there is little theoretical or modeling guidance on how one might anticipate risks without detailed empirical work. They concluded, "As of yet, we are unable to predict which successful invasions may have the biggest impacts." Conference participants discussed the use of experimental versus "natural world" science in the case of phytosanitary issues. For example, one commentator argued that ecologists are less willing to extrapolate calculations because part of the business of ecologists is to emphasize variability and unpredictability in the natural world.

One of the problems of predicting the biological impacts of invaders is that the organisms themselves are often not well known. Indeed, as Goodell et al. pointed out, the basic taxonomy—simply identifying what species is present—is commonly a limiting factor. But even organisms that seem "well known" have their surprises. For example, if one is performing a risk analysis on a well-studied crop plant that has been carefully engineered using recombinant DNA technology, it might be expected that the risk assessment would be a simple matter. Moreover, as a prelude to actually commercializing a transgenic plant (a crop with DNA from another species inserted using recombinant DNA technology), hundreds of greenhouse and field trials are typically performed. However, as Peter Kareiva and Michelle Marvier pointed out in their review of different international approaches to transgenic crop regulations, surprises occur with these well-studied plants and the uncertainties of ecology and evolution confound evaluations of even the most domesticated organisms (Chapter 10).

The sophistication of scientific approaches to GMOs has increased enormously in the past few years. Initially, "safety" was naively assumed to be potentially ensured by strategies for containment, and intrinsic risk was thought to be characterized simply by listing the traits of the crop plant being modified, without any primary data. Currently, a wide variety of national policies regarding GMOs in agriculture have converged on some common approaches: (1) it is the phenotype of the modified plant that is key, not where the DNA came from; (2) familiarity offers some safety assurance; and (3) if the traits are likely to confer some ecological advantage to a recipient plant, then data regarding their transfer to wild relatives and the likely impact in a wild population are desirable. However, there are again no "model systems" for making risk assessments or any well-accepted models or theories that might lead to quantitative estimates of risks, or even "bounds on risk." In GMOs, there is increasing emphasis placed on monitoring as a device for catching any mistakes, including any cases where an unsafe transgenic plant is allowed to be commercially produced.

John Stark's review of ecotoxicology opened an entirely different window on ecological risk analysis, but with familiar results (Chapter 3). Human toxicology uses well-established laboratory animals for toxicity studies and has well-defined standards. Ecotoxicology initially adopted a similar approach, producing dose-response curves for standard organisms such as honeybees and Daphnia. As ecologists have become more involved in the field of ecotoxicology, they have criticized this simple model system approach and have sought more ecologically meaningful measures of risk. Stark pointed out that it is not obvious which species should be used when assaying chemical toxicity, and because we obviously cannot test all species, some decision must be made. Beyond the difficulties of selecting a species, Stark also pointed out that traditional mortality studies miss the important fact that toxicity strongly depends on population structure (e.g., relative numbers of juveniles versus adult individuals). Thus, as ecotoxicology becomes more realistically "ecological," its risk assessments lose the false security of simple models systems and straightforward quantitative tools, such as dose-response curves.

If one were to try to reduce the above four case studies and presentations of ecological systems and their risk analysis issues to one message, one might be tempted to conclude that ecological risk analysis is anecdotal and haunted by "special cases" as opposed to general principles: There are no standard calculations and no model systems or even reference systems. These judgments partially reflect the fact that ecology is profoundly influenced by the details of species' associations and historical events. The processes also include many nonlinear indirect effects. In this context an indirect effect can be as simple as the observation that if a pesticide kills pest insects and their predators at equal rates, the net result will be higher prey populations. This "indirect effect" arises because mortality on predators does more than remove predators; it removes a negative feedback on pest population growth. Indirect effects can get much more complicated. For example, it would have been almost impossible to predict that gypsy moths enhance the prevalence of Lyme disease, as reported by Goodell et al. (Chapter 5).

Scientists have called for greater investment in databases for ecotoxicology, non-indigenous species, pests, and GMOs. In the absence of good models, the best route may be an encyclopedia of examples that can be consulted for statistical generality. In general, this approach is likely to become a "principle" of ecological risk assessment—generalities will have to be empirically rather than theoretically based. To some extent, this view is consistent with the use of meta-analysis to combine diverse empirical evidence from experimental and nonexperimental sources and recover insights into the underlying processes. The purpose of statistical analyses can be used to detect all influential factors, including small and subtle effects that could become net large impacts with widespread introductions. However, the very acceptance of such an empirical and statistical "model" carries with it the implicit assumption that mistakes may be made. It is no accident that there is also a trend within ecological risk assessment toward the requirement of large monitoring programs. What this means with respect to phytosanitary standards is a foundation that is largely statistical and empirical and that includes a commitment to funding for the support of databases and monitoring programs.

Although the above discussion focuses on challenges in ecological risk assessment, risk assessment also faces significant challenges in other areas of SPS management. For example, the case study presentation by Bent Nielsen illustrates the level of analytical and monitoring resources necessary to track sources of *Salmonella* in pig production and pork processing (Chapter 8). In all areas, the international standard-setting bodies such as *Codex*, OIE, and IPPC, as well as bilateral and multilateral cooperation, play an important role in developing protocols necessary for sound risk assessment. An outstanding question is the relationship between sound risk assessment protocols for different SPS risk sources. This is particularly important because the SPS Agreement includes consistency of treatment across risk sources as a criterion for evaluating SPS measures.

The discussion identified additional points regarding risk assessment. Governments recognize that the risk assessments they present in support of their SPS measures have the potential for establishing precedents regarding what is required for a sound risk assessment. Therefore, there is a strategic element in the risk assessment process that influences the factors considered and the quantity and quality of evidence provided. Paul Thompson's case study discussion of GMOs pointed out that the risk assessment community is often quite sophisticated in their deliberations of what is important and how to measure it, whereas citizens/consumers are at an earlier and more basic stage (Chapter 10). The latter may be just beginning to consider whether a risk should be allowed or eliminated and who should be accountable for it. Thus, risk assessment has to involve risk communication from the analysts to citizens/consumers and vice versa. Thompson noted that it would be irrational to deliberate over everything. Too much information creates congestion. We need some sorting mechanism to ensure that important issues are identified and that there is sufficient discussion of the risk management issues they pose. Several participants rejected the notion that if scientists are left to get on with their work, they will be able to deliver objective risk assessments that in turn could be effectively handled by risk managers and communicators as a technocratic process. According to many participants, this is folly. There is a clear need for an iterative risk assessment, management, and communication process.

Economics: Measuring the Costs and Benefits of SPS Management Strategies

A third theme of the conference was the role economic analysis plays in the design of SPS measures and in their international management. In their paper (Chapter 7), Jean-Christophe Bureau and Stephan Marette argued that "The idea of objective science serving to guide trade practice, which prevails in the SPS Agreement, is debatable. In practice, economic and political considerations are very much intermingled." Most fundamentally, economic considerations influence the risks that are judged as important for assessment. Risk assessment and management require resources. As a consequence, there is an allocation task among competing alternatives and between risk and non-risk-related activities. There will never be sufficient resources to consider all. Thus, only those risks that are important enough are addressed. The economic costs of risks, and the potential benefits of controlling or reducing them, are key factors determining which risks get attention. Risk management focuses on the ways risk may be reduced to an acceptable level, which includes economic considerations. Finally, economics plays a role in determining which disagreements over SPS measures merit the bureaucratic resources required for consultations between countries and, ultimately, WTO disputes.

Economic analysis was seen as contributing to the management of SPS issues in several ways. Preliminary benefit or cost analyses can identify the priority SPS issues to be addressed. Cost–benefit analysis also provides an economic measure of the impact of particular SPS regulations. This allows risk managers to identify the most effective strategies and also to gauge whether a proposed measure meets the criterion of the SPS Agreement that it be "least trade restrictive." Although international agreements do not oblige countries to adopt only those regulations whose benefits exceed their costs, analysis of this type may help to avoid SPS measures that clearly decrease welfare and in recognizing the distribution of their benefits and costs within and between countries. In the United States, cost–benefit analysis is institutionalized as a part of regulatory decision making by the requirement that such an analysis accompany major regulations. Furthermore, economics provides analysis of the "market" for SPS protection and of the incentives for governments, companies, and consumers to provide such protection.

More generally, economic approaches address the welfare outcomes of SPS and other regulatory measures. Bureau and Marette argued that cost-benefit analysis should take a more prominent place in analyzing trade policy in the SPS area, especially for food-related issues where consumers are interested in a broad range of food attributes beyond safety, including process attributes and ethical and cultural considerations. Estimates of consumers' willingness to pay for particular food characteristics may help to clarify their level of concern about those characteristics.

Dispute settlements to date have put risk assessment ahead of other types of analysis. If economic methods were used more systematically, the welfare gains resulting from specific SPS measures could be compared with the welfare gains resulting from freer trade. For example, if a WTO-consistent SPS regulation results in the import of products that do not satisfy consumers' safety, ethical, environmental, or cultural concerns, consumers may avoid those and similar products. The resulting market disruption could lead to a substantial welfare loss. Bureau and Marette noted that it would be paradoxical if trade liberalization, accompanied by international rules designed to settle SPS disputes, were to result in more trade but less welfare. However, several conference participants cautioned against relying too heavily on economic approaches to measuring what is important. One participant said that cost–benefit analyses of trade barriers have the effect of marginalizing the concerns of many people.

David Victor made a point related to Bureau and Marette's paper regarding trade and welfare impacts in his discussion of the three disputes on SPS measures that have been decided by the WTO Appellate Body (Chapter 6). Although the decisions made extensive use of risk assessments, they also set a standard that the SPS measures adopted have a "rational relationship" to the assessed risks. This in part requires an analysis of how the measures affect trade. He concluded that, as a growing number of national measures come under scrutiny for their consistency with the SPS Agreement, a requirement for "trade impact assessment" will probably become commonplace.

Political Science: Establishing International Discipline while Preserving National Sovereignty

A fourth theme expressed throughout the conference was the tension between establishing an international discipline on SPS measures as they affect trade and the preservation of countries' abilities to deliver the level of SPS protection that they and their citizens desire. We are seeing an international legal system in the making, with early cooperative efforts among countries, activities of the international standard-setting bodies, and WTO disputes shaping perceptions about how well the system is working.

David Victor discussed the experience to date with the SPS Agreement, particularly in the context of the three cases that have gone through to the WTO Appellate Body level of the dispute settlement process. Although the SPS Agreement gives more prominence to international standard-setting bodies, in the early period their role was not crucial because decisions rested on the quality of the risk assessment rather than on whether it was in conformity with an international standard. In fact, Victor argued that standardization of approaches to and the use of risk assessment, rather than of SPS standards themselves, is the most likely outcome under the SPS Agreement. It is even possible that the agreement could result in more diversity of standards as better risk assessments are done and more information is utilized, or as countries bring less stringent standards into line with stricter ones. Victor argued that the fact that the cases found the challenged SPS measures to be illegitimate does not indicate that the agreement is biased against strict SPS regulation. Rather, they involved measures that were readily established as not based on sound risk assessment. Moreover, they did not set clear standards for judging whether an adopted SPS measure is "least trade restrictive."

As a result of the SPS Agreement, countries are becoming more disciplined and internally consistent in risk assessment and management. This evolution in discipline is part of the procedural standardization discussed in Victor's paper (Chapter 6). It will likely promote transparency and facilitate determinations of equivalency between different countries' standards. The SPS Agreement will likely always have "teeth" due to the requirement of a risk assessment. International standards and risk assessment protocols may facilitate SPS management but are not crucial to the successful operation of the SPS Agreement. Victor viewed the SPS Agreement as being more accommodating to other than strictly scientific concerns than it is sometimes portrayed. He said that the appellate decisions in the three cases have actually given more latitude to countries than the writers of the SPS Agreement envisioned, but that is probably politically wise.

In discussing the Hass avocado case study, David Vogel argued that NAFTA, like the SPS Agreement, places a desirable discipline on political choices within a country (see Chapter 9). Prior to NAFTA, the U.S. government responded with a full quarantine to the potential risk of fruit fly infestation as a result of avocado imports from Mexico. After NAFTA came into effect, a new risk assessment was conducted, and the United States adopted a policy allowing imports under certain conditions and with specific controls. The new policy facilitates trade, but still allows the United States to achieve its desired level of protection. Vogel noted that this policy choice was not politically feasible without the outside discipline of NAFTA. Trade agreements can formalize and enforce reciprocal arrangements that are beneficial to trading partners, while limiting the ability of specific interest groups to influence policy in their own favor. However, the Hass avocado case was relatively straightforward because it did not involve issues of human safety. The ability of science alone, based on risk assessment, to work out a solution was greater in this situation.

A continuing theme in the conference discussions was the issue of power in SPS decision making. Questions focused on who participates in international deliberations and how much power large companies or nongovernmental organizations have in these discussions. Who is not represented in the discussions? Who are the primary beneficiaries of the current system and who ends up paying the cost of mistakes? A related issue discussed was representation and participation of different countries in the international decision-making process. There was recognition among several participants that the countries most active in developing international SPS institutions represented the interests of well-to-do citizens in wealthy countries.

Much conference discussion focused on how different countries and citizens/consumers view the evolving balance between international discipline and national sovereignty. One important determinant of this view is whether the new international institutions are perceived to be resulting in an increase or decrease in domestic levels of SPS protection. David Vogel asked whether the new institutions will create incentives for countries with lower standards to move into line with those with higher standards (leveling or harmonization up) or vice versa (leveling or harmonization down). In the area of meat slaughtering and processing standards, Bruce Silverglade saw evidence of a leveling down effect (see Chapter 8). He saw the United States–European Union beef hormone dispute as an example in which the United States, supported by the WTO, is attempting to limit the European Union's choice of a zero-risk standard. He viewed the acceptance of company employees conducting food safety inspections for meat products to be exported to the United States as another example of leveling down.

David Victor saw little evidence of a race to the bottom in the three SPS disputes decided to date. But he did see a political risk that, in transferring attention to the international system of law in this area, there could be a backlash, especially if the system is not flexible enough. Mandatory compliance to ill-accepted international norms may result in citizen/consumer rejection of freer trade, which already has a poor standing in public opinion in many countries. Bureau and Marette argued that food, especially, is a sensitive topic, and few things are more likely to call trade liberalization into question than to have it associated with foisting mediocre, undesirable, or even potentially unsafe products on consumers. The prospect for this type of response is increasing, especially in Europe and parts of Asia, as consumers place increased emphasis on the cultural, ethical, and environmental attributes of agricultural and food products. These attributes may not fit the constructs of the SPS Agreement, or they may fall between the cracks of the SPS and TBT agreements.

The emphasis of the SPS Agreement on "science" may conflict with the application of the precautionary principle and recognition of other legitimate factors in SPS decision making. Both concepts are under development as applied to SPS issues. The precautionary principle addresses how to proceed (i.e., how much precaution to exercise) when information for risk analysis is inadequate. A statement of the precautionary principle widely used in international discussions is offered by Principle 15 of the Rio Declaration on the Environment and Development: "Where there are threats of serious or irreversible damage, lack of full scientific certainty shall not be used as a reason for postponing cost-effective measures to prevent environmental degradation." Approaches to implementing the precautionary principle in the SPS arena are being discussed. Other legitimate factors refer to additional considerations (e.g., economic development, preservation of traditional production practices) that governments may wish to incorporate in their decision-making. These two concepts are increasingly referenced by governments and public interest groups as desirable additional bases for SPS decisions.

At the same time, it was clearly recognized in the conference discussion that safety and other concerns give trade protectionists an opportunity to cheat on market access; cultural and ethical arguments can be used to cover a potentially unlimited number of exceptions to free trade. The discussion focused on the need for an evolving approach to the management of SPS issues in international trade that balances the need for discipline in market access with some safety valves that recognize countries' own desires and, in some cases, their needs for transition time. There was concern that the SPS Agreement's safety valves (recognition of countries' rights to choose the appropriate level of protection, the ability under the agreement for countries to compensate those who are damaged by their disputed standards rather than change the standards, and the ability to adopt interim standards where scientific evidence is not yet available) may prove inadequate. If they do not meet the needs for flexibility, then support for market access and freer trade will diminish.

SUMMARY

The conference presentations and discussions focused on recent experience with the management of SPS issues in international trade. This discussion necessarily paid great attention to the provisions of the SPS Agreement of the WTO. Nonetheless, it was recognized that this agreement is part of a much more extensive international effort to balance market access and free trade with countries' abilities to provide SPS protection within their boundaries. The discussion among conference participants can be summarized in the following points associated with the roles for science, culture, politics, and economics in the design of SPS regulations (measures) and their international implementation:

• The SPS Agreement, negotiated as part of the Uruguay Round of GATT, is the central current framework for SPS management in international trade. It uses risk assessment as a "scientific" response to the need to assess existing SPS and related technical measures as barriers to trade. It also incorporates other criteria, for example, through the provision that SPS measures should be the least trade restrictive possible.

• Domestic and international experience suggests that a comprehensive approach to risk issues is important. In recent years such an approach has been followed within the framework called risk analysis, which includes risk assessment, risk management, and risk communication. This experience also suggests that risk analysis must consider the multidimensional nature of the framing of risk issues. Risk analysis requires the input of many of the sciences, including the biological and natural sciences, as well as social and behavioral sciences and the law, although each may be more prominent in different phases of the risk analysis.

• Development of consistent protocols, within and across countries, for using risk analysis and resolving SPS issues is an evolving process. The SPS Agreement itself does not explicitly use the full framework of risk analysis but contains elements of it. The initial emphasis in disputes has been on risk assessment, but decisions in those cases, as well as experience in bilateral and multilateral negotiations on equivalence, indicate that risk management and communication issues are also central to addressing SPS issues.

• Resolution of differences in approaches to risk, including risk assessments, may lead to greater use of comparative risk evaluations. For example, a country's risk standard for one SPS threat may be evaluated relative to how it tolerates and manages risk for another SPS threat. Comparative risk evaluation is also key to determining equivalence in countries' SPS measures and may contribute to harmonization of regulatory approaches.

• Management of SPS risks requires specific guidelines for risk analysis with input from the biological and natural sciences, economics, sociology, political science, and other disciplines. These guidelines may outline the various phases of risk analysis and indicate where these different approaches make their most direct contributions. They could also provide a clear discipline on SPS measures that function primarily as barriers to trade.

• There may be merit in exploring a role for biological, natural, and social scientists in providing input to the WTO on developing criteria documents on oversight of risk analysis practices, development of approaches to comparative risks, and coordination of risk analysis with cost-benefit analysis.

• As trading partners seek to establish a discipline on SPS measures and coordinate their policies, it is important to build some safety valves into the process in order to recognize marked differences in regulations due to differences across countries in the evaluation of risk sources and events at risk. Researchers and policymakers could consider the characteristics of safety valves, including compensation, that when built into the system may smooth establishment of reasonable SPS discipline, maintain support for freer international trade, and reasonably protect countries' abilities to manage SPS risks.

Prospects for the Future

It is important to recognize that a great deal of progress has been made in the management of SPS issues in international trade. The new trade agreements and bilateral and multilateral efforts have brought a desirable discipline to national decision making on SPS measures. All the sciences have contributed to this progress. The social sciences, for example, have contributed through institutional design. The biological and natural sciences have contributed through the development of protocols for risk assessment. It is important to recognize that the prominent trade disputes before the WTO are only a small part of the process, with multilateral and bilateral cooperation writing the larger story. However, the disputes are important beyond the trade and SPS issues affected because they form perceptions about the costs and benefits of the new trade discipline.

As we move forward, development of a systematic framework addressing all the important factors that influence SPS decision making and its impact on international trade may be desirable to improve market access and address the backlog of SPS concerns. We have such a framework in the form of the risk analysis paradigm. This framework highlights the roles that the different sciences can play in the management of SPS issues in international trade. Politics, sociology, culture, and economics play a role in determining which risks will be subjected to a risk analysis. In effect, there has to be something striking about a particular risk before resources are expended to understand and analyze it. The risk assessment phase of risk analysis relies heavily on the biological and natural sciences. This is what the SPS Agreement primarily refers to when it discusses "science." But the social sciences and philosophy also play a role here because risk assessment in part rests on predictions of how humans will act. Politics, economics, and culture play prominent roles in risk management in the choice of measures to address important problems identified in risk assessments. The SPS Agreement requires that the measures adopted be the "least trade restrictive." Economics can play a strong role in measuring costs and benefits of SPS measures and evaluating their level of trade restrictiveness. The natural sciences can play an important role in risk management as well, particularly in identifying effective risk management strategies and approaches. Finally, the social sciences are prominently featured in risk communication.

A key consideration for future progress in the management of SPS issues in international trade is the recognition that risk analysis is an iterative process. Risk assessment is not followed in lock step by risk management and then risk communication. Managers move back and forth between the parts of the risk analysis using the different sciences, the cultural frameworks of the respective countries, and domestic and international political and economic considerations in making their decisions. Future progress can be made in the management of international SPS issues through recognition of the competencies and contributions of the different sciences and their systematic and integrated use in risk analysis. Ultimately for our international systems to work effectively, they have to be viewed as working effectively by citizens/consumers, governments, and businesses in countries around the world. Clearer integration of a broad range of sciences and recognition of different viewpoints may be key elements in meeting this test over time.

1

Historical and Social Science Perspectives on the Role of Risk Assessment and Science in Protecting the Domestic Economy: Some Background

G. EDWARD SCHUH

Humphrey Institute of Public Affairs, University of Minnesota

There is hardly any policy problem on this nation's agenda that does not require the collaboration of the social sciences, the biological and natural sciences, and engineering to devise suitable policy choices for society. The design of policy is essentially the design of institutional arrangements for our society. That cannot be done effectively without the participation of multiple disciplines.

Although the need for multidisciplinary collaboration might be obvious once one gives the problem a moment's thought, it is not an easy objective to achieve. Differing perspectives on important problems make it easy to fall into disciplinary stereotyping and quarreling that can be quite counterproductive. It takes a special effort to understand the specialized language we all tend to use and to understand the perspectives offered by disciplines other than our own. Yet both efforts are needed if we are to capitalize on the insights offered by various disciplines. The search for understanding the perspectives of other disciplines can in particular be hard intellectual work.

Both our students and the public deserve the best we can give them from multidisciplinary collaboration. Students who can observe and participate in multidisciplinary endeavors will develop open and inquiring minds. On policy issues, the public will gain insight into the complex and complicated world in which policy is shaped. Moreover, our citizenry will benefit from the best that science can offer in policy alternatives from which to choose. In effect, although a multidisciplinary approach takes time and effort because of the need to understand each other's language and alternative perspectives, our social product will tend to be higher if we take this approach.

HISTORICAL PERSPECTIVE ON PROTECTIONISM

Before providing a historical perspective on protectionism, it is useful to first provide a brief outline of the benefits of international trade. Economists tend to be the ones who articulate the benefits of international trade. These benefits tend to be subtle and not always directly observable, whereas the costs in terms of displaced workers and perceived effects on the distribution of income are more explicit and observable.

The easily observable benefits of international trade are the expansion of consumer choices and the reduction in prices of goods and services it brings. An important example of this occurred when the United States was flooded with imports of automobiles from Japan. The United States was fortunate that protective measures against imports at the time were quite low. U.S. automobiles were not keeping up with the quality of automobiles from Japan and Germany, and they were expensive in comparison. Imports expanded quality opportunities for U.S. consumers, and at a lower price. That is the means by which increases in per capita incomes come about.

The experience with Japanese automobiles teaches us another lesson. Recent developments in economic theory have linked international trade to economic development. Trade is not only an important means of bringing new technology into a country, but it forces domestic economies to modernize and thus lower their cost of production, again benefiting consumers with declines in the price of consumer goods.

In recalling what happened in the case of automobile imports from Japan, it did not take many years for U.S. automobile companies to catch up on the technology that the U.S. consumers wanted imbedded in their automobiles. Moreover, in a relatively short period of time they were also able to modernize their production practices, so they became more competitive with imports. Ultimately, jobs were saved here at home, but in a way that contributed to economic growth and development.

This points to another aspect of international trade worth noting. Economic growth benefits from an expansion of trade in part because the foreign exchange increased exports earn can be the means of financing a higher rate of economic growth and development. But the benefits will tend to be much broader. International trade enables individuals to benefit from an international division of labor that is based on specialization and on the unequal distribution of resources around the world. By making more efficient use of the world's scarce resources, everyone participating in that division of labor has their standard of living raised. The exceptions, of course, are those most damaged directly by the trade, which will be discussed below.

In regard to the historical perspective on protectionism, if one goes back in time, tariffs were the primary barriers to trade. Moreover, many, if not most, tariffs were imposed as a means of generating revenue to finance the government. Protection as we understand it today was not the main objective.

During the Great Depression of the 1930s, tariffs grew in importance as a protectionist measure. They became large and pervasive. Almost every nation was using them for that purpose. When the General Agreement on Tariffs and Trade (GATT) was created at the end of World War II, its main focus was on the reduction and eventual elimination of tariffs, and specifically on the reduction of tariffs on manufactured products.

As the GATT became successful in lowering that kind of protectionist barrier to trade, however, another form grew in importance—nontariff barriers. Nontariff barriers to trade take many forms, one of which is sanitary and phytosanitary (SPS) barriers.

Although SPS standards may be justified in some cases, they tend to be an important nontariff barrier to trade in many other cases. This particular form of trade protectionism is especially pernicious because of the lack of transparency in their implementation. They are not visible to consumers, and political support for them can be mustered with little opposition. Moreover, it is difficult for consumers to understand the technical issues even when they are transparent. For example, prior to the North American Free Trade Agreement, the United States had a nontariff barrier on imports of fruits and vegetables from Mexico to the effect that any trace of soil on the imports was grounds for a barrier to importation. How were consumers to know that the issue was whether "night soil" (human waste) had been used in the production of the commodity rather than the presence of soil per se? An important issue to be sorted out in examining trade policy is whether the standards to be considered are a garden variety of economic protectionism or whether there are more serious issues at root.

In today's world the protectionists tend to hang their arguments on two sets of issues. The first is food safety, in part from new biotechnology innovations. This concern about food safety has come in a number of forms, ranging from the bovine somatotropin (BST) hormone in milk, to *Escherichia coli*, to mad cow disease, to the use of hormones in the feed for beef cattle, to irradiation of food. The food safety issue has arisen in part because of the globalization of sources of food supplies through international trade and, in part, because of rapid technological progress, which is producing innovations in the production and processing of food. Both of these pose real threats to food safety.

The second issue is the perception that international trade is causing the distribution of income to become more highly skewed or unequal as trade expands. This issue deserves some attention because the failure to deal with the problem of labor adjustment is often what drives the search for nontariff barriers as a means of protection. At the same time, however, more attention could usefully be given to what happens to the absolute income of the poor as trade is liberalized. This is a rather neglected issue, despite its importance.

An important means of dealing with the food safety issue is to be able to make a proper assessment of the risk from these external food supplies and the design of policies to protect the consumer from dangers that are potentially inherent in such products.

A PERSPECTIVE ON RISK ASSESSMENT

Although the collaboration of biological and natural scientists with social scientists is important in almost all dimensions of international trade policy, it becomes especially important in making risk assessments and designing risk management strategies. The biological and natural scientists have important contributions to make in assessing the technical dimensions of the risks, and social scientists need to take their assessments into account. The social scientists contribute important insights when they estimate costs and benefits of the innovations or risks, and the biological and natural scientists need to understand the social scientists' perspectives. One important issue in making risk assessments and in designing policies to manage that risk is that there appear to be cultural differences in perspectives toward risk. For example, Europeans seem to be much more averse to some food safety issues, such as hormones, than are Americans. Americans, on the other hand, are more averse to carcinogens. If these differences are real and significant, the next issue is whether international trade regulations should take such differences into account. Moreover, there is the issue of how they should be taken into account.

An important empirical issue in this context is that in the post-World War II period, the Europeans have been much more protectionist toward their agriculture than has the United States. The issue thus becomes whether what is perceived as a food safety issue is not just a subterfuge for plain old economic protectionism.

There is a great deal in the literature to assure us that different people have different tastes for risks. A study done by the North Central Farm Management Committee in the 1950s showed that beef producers, for example, tended to have a greater taste for economic risk than producers of commodities such as the grains, for which price stabilizing government programs were in effect (Halter, 1961). Similarly, common observation tells us that many people are willing to gamble at unfair odds, which means that they are willing to take the risk even though they know the chances are good that they will lose. On the other end of the perspective, we know that some people are willing to insure against some kinds of losses, whereas other people will not insure against the same losses.

Professors Friedman and Savage clarified some of these issues many years ago. They showed, for example, that it would be rational behavior for people to gamble at unfair odds for possible gains that would change their socioeconomic status. Similarly, they would pay to insure against losses that would lower their socioeconomic status (Friedman and Savage, 1948).

An important issue is whether the rules of trade need to take account of the risks to the food supply posed by new technology in the production and distribution system, and if they do, then how should these risks be taken into account? That is, how should the rules be structured and defined?

Much of the literature on risk assessment and management is based on microdecisions involving such things as the introduction of a new drug or the establishment of a power plant—interventions that affect a limited number of people and under generally unique or special conditions. An intervention in trade policy, however, usually affects a much larger number of people, with a wide variety of individual circumstances.

How should we proceed under these circumstances? U.S. policy on these issues has for the most part argued that the trade rules should be science based. I do not believe this argument goes far enough. The role of the biological and natural scientists, and of engineers, is also to help establish the probability distribution of the risks. They also have a contribution to make in understanding the consequences of the threat to food safety. Whether the consequence be death, a serious health problem, or something relatively mild, the information is crucial.

The general tendency is to think that biological and natural scientists should be involved in quantifying such risks because they are objective in evaluating the risks. And within a wide range of circumstances they are. However, we need to recognize the limitations of this objectivity. For example, the very way a question is framed may introduce bias into the analysis, perhaps by focusing the question too narrowly for policy purposes. Similarly, even the assumptions behind the statistical analysis can lead to bias in the analysis. For example, the assumption of the null hypothesis as used in risk analysis contains an implicit bias because it places a greater burden of proof on those who would avoid or limit a hazardous activity, presuming these activities are safe until proven otherwise.

There is also the point that, in the final analysis, science can never be an adequate basis for a risk decision. Risk decisions are ultimately public policy choices. In the case of food safety, these choices ultimately depend on the trade-offs among the groups that are affected and others in society, the value of life and interpersonal comparisons of utility. Unfortunately, losses that one individual experiences can be compared with the benefits others realize only under rather limited circumstances. That is what ultimately brings the political process into play.

Despite these caveats, the calculation of the technical risks is the logical starting place to define policy and new institutional arrangements. The next step is to determine the costs and benefits of the particular intervention under consideration. It is easy for an economist to believe that is his or her bailiwick. Obviously, they should have many of the technical skills to make such an analysis. However, the biological and natural scientists, and other social scientists, have much to contribute to identifying who benefits and who pays the costs.

Still another part of the analysis is to know something about the distribution of those benefits and costs, which of course, will ultimately influence the significance of the benefits and costs. Are just a few people going to be affected? Or will either or both the benefits and costs be widely distributed in society? That may influence the nature of the rules ultimately established. Sociologists have much to contribute to the analysis of the distributional issue.

Unfortunately, these distributional issues have, for the most part, been ignored in the past. The emerging environmental justice movement in the United States is beginning to give them more attention. More detail on environmental justice can be found in the Institute of Medicine publication, *Toward Environmental Justice: Research, Education, and Health Policy Needs* (1999). The important problem for the design of policy interventions is that technically we know very little about how to compare the costs and benefits across members of society. That is where political scientists come into play. They should be able to assist in designing policies that are politically acceptable and enforceable.

Another important factor is the role of culture in the design of policies. Such things as the desire for individual freedom, the role the citizenry see for government in their society, and related issues must be understood. Americans, for example, tend to elevate individual freedom to a high value. Europeans, on the other hand, often seem to be less concerned with individual freedom and tend to think more in terms of the common good. Similarly, different people have different tastes for policy mechanisms. Europeans seem to have little distaste for taxes, and in the United States we seem to have a serious distaste for them.

There are two related issues. The first is the role of lay people in the assessment of risks and in the design of policy. Ultimately, the design and acceptance of policy is a political issue. Experts have much to offer on the technical side, but they by no means have a monopoly on knowledge. Lay people can help force the discussion onto relevant issues, and they often bring specialized knowledge to the table. (The 1996 National Research Council's report, *Understanding Risk: Informing Decisions in a Democratic Society*, calls attention to the value of lay knowledge.) The involvement of lay people takes time, however.

Finally, there is the importance of information and of informing the body politic on the issues. Transparency in putting the appropriate knowledge before the body politic is an imperative in a democratic society. Sound policy—policy consistent with the desires of society—can be obtained only if the citizens are informed on the issues.

The importance of this issue can be seen by a recent newspaper article on the BST hormone and milk. The article reported that some 85 percent of the milk sold in the Twin Cities area (Minneapolis/St.Paul, Minnesota) was now produced by cows being treated with this hormone. The article marveled at the ease with which the hormone became widely used given the controversy when it was first discussed. The question that begs asking, however, is how many consumers knew whether their milk was actually produced with that treatment or recalled the controversy surrounding its use?

THE IMPORTANCE OF ADJUSTMENT POLICIES

Despite all the benefits associated with international trade, some groups in society are often disadvantaged by it. The more aware and active among these groups are the ones who lobby for protectionist interventions, and these interventions often take the form of SPS standards.

To the extent that such standards are motivated by protectionist aims, the key to addressing them is to alleviate the pressure for intervention. This can best be done by adjustment policies that help those who are disadvantaged to shift to alternative employment—whether it be members of the labor force or owners of private capital.

The structural and sectoral adjustment loans implemented by the World Bank and the International Monetary Fund in the 1980s are examples of attempts to deal with these problems at the international level. The structural adjustment loans tended to provide balance of payments support while changes in policy were implemented, thus cushioning the stress that is inevitable in policy reform. The sectoral adjustment loans were designed to facilitate the reallocation of resources to the tradable goods sectors, thus helping to accelerate the adjustment process and by that means reduce the pain and suffering entailed in reorganizing the economy implicit in the adjustment process.

These adjustment policies could have been made more humane by the judicious use of food aid to support targeted feeding programs directed to low-income urban consumers and workers. These are the groups who tend to be most disadvantaged by the rise in food prices associated with the realignment of currency values and other reforms that were such an important part of the policy reforms.

A remaining challenge is to design domestic labor adjustment policies that are politically acceptable and which help displaced workers shift to alternative employment. Where geographic mobility is required to gain alternative employment the financial and psychic costs can be important, to say nothing of the need for new skills. Unfortunately, domestic political leaders are seldom in favor of losing their political base through out-migration and thus tend not to favor such policies.

CONCLUSIONS

There are two important aspects of designing policies that we should not lose sight of as we continue with our work into the future. The first is that we can expect the globalization of our economy to continue to grow apace in the future. It has been rooted in rather basic and far-reaching technological revolutions in the transportation, communication, and computer sectors of our economy that have dramatically lowered the cost of international trade and substantially increased the scope of markets. Moreover, these revolutions have hardly yet reached the previously centrally planned economies, or the developing countries where 80 percent of the world's population live.

Second, the key to sound policy in the future is for all of us to be doing the analytical research needed as the basis of policy design. This analytical work will be no better than the multidisciplinary collaboration that we are able to develop and our willingness to involve lay people in our discussions at all levels.

REFERENCES

Friedman, M. and L.J. Savage. 1948. The Utility Analysis of Choices Involving Risk. The Journal of Political Economy. August. Vol. LVI(4):279–304.

Halter, A. 1961. Utility of Gains and Losses. Pp. 128–139 in G.L. Johnson, A. Halter, H. Jensen, and D.W. Thomas, eds. A Study of Managerial Processes of Midwestern Farms. Ames: Iowa State University Press.

IOM (Institute of Medicine). 1999. Toward Environmental Justice: Research, Education, and Health Policy Needs. Washington, D.C.: National Academy Press.

NRC (National Research Council). 1996. Understanding Risk: Informing Decisions in a Democratic Society. Washington, D.C.: National Academy Press.

Part I

Agricultural Trade, Risk Assessment, and the Role of Culture in Risk Management

2

Sanitary and Phytosanitary Risk Management in the Post-Uruguay Round Era: An Economic Perspective

DONNA ROBERTS[1]

Economic Research Service, U.S. Department of Agriculture,
U.S. Trade Representative's Mission to the World Trade Organization,
Geneva, Switzerland

Although many governments are now committed to reducing the number and rigidity of regulations that are thought to stifle economic innovation and competition, it is widely expected that the regulatory environment for agricultural producers and processors will become more complex in the coming years (Organization for Economic Cooperation and Development [OECD], 1997). Income growth is fueling demand for environmental amenities and food safety, and increasingly regulators are being asked to provide these services when markets fail to do so. On the "supply side" of regulatory activity, U.S. officials who devise sanitary and phytosanitary (SPS) measures—regulations that sometimes restrict imports in order to reduce risks to animal, plant, and human health—face additional challenges. These officials are now bound by the multilateral legal obligations found in the Agreement on the Application of Sanitary and Phytosanitary Measures (SPS Agreement; see Appendix A) of the World Trade Organization (WTO) which came into force in January 1995. Moreover, recent regulatory reform initiatives, including Executive Order (EO) 12866 and other Executive Branch directives, revise previous guidelines for

---

[1]The author wishes to thank David Orden and Julie Caswell for their comments on this paper. The views expressed in this paper are those of the author and do not necessarily represent the positions of U.S. Department of Agriculture or the Office of the U.S. Trade Representative.

basing decisions about major regulations on assessments of their benefits and costs. Taken together, these developments have substantially changed the parameters for regulating imports of agricultural products from the time when "when in doubt, keep it out" was viewed as an appropriate decision rule.

It is clear that the domestic regulatory reform initiatives share many goals with the SPS Agreement. For example, both advocate transparency of regulatory rule making in order to promote symmetry of information among stakeholders, which include agricultural producers, processors, and consumers on one hand and trading partners on the other. Both also require that a regulation be based on a careful assessment of the risks that the measure is designed to mitigate and make provision for the inclusion of the costs of control programs as a factor in regulatory decisions.

However, in other respects, it is unclear whether the legal obligations found in the SPS Agreement are wholly congruent with U.S. Executive Branch guidelines for consideration of economic efficiency and distributional effects of measures as decision criteria. The SPS Agreement is primarily intended to aid WTO members in the decentralized policing of regulatory protectionism in foreign markets. Regulatory protectionism or capture occurs when domestic groups with a vested interest in a particular regulatory outcome successfully lobby for overly restrictive SPS measures that, by limiting or preventing safe imports, lower net social welfare. Two requirements in the SPS Agreement—to base SPS decisions on a risk assessment and to notify trading partners of changes in SPS measures—underpin the multilateral monitoring system. The risk assessment paradigm of the SPS Agreement, centered on the concept of "acceptable level of risk" (or "appropriate level of protection" in the language of the agreement), endorses risk-related costs as a normative basis for SPS regulatory decision making. This concept implicitly excludes consideration of benefits to other economic agents and generally fuses risk assessment with risk management by embedding value judgments about which risks are "acceptable" into scientific assessments. This approach stands in contrast to the economic paradigm of the Executive Branch directives in which normative rules for designing SPS measures rest on cost–benefit analysis tools to *infer* appropriate levels of protection from individual preferences.[2]

The simultaneous emergence of new multilateral and domestic rules for SPS regulatory decision making highlights the need for a comprehensive examination of this new regulatory environment. In this chapter I review the SPS Agreement with a view to examining how the agreement does or does not constrain the use of economic analysis in the design of regulations for imports

---

[2]EO 12866 (1993) requires agencies to perform a cost-benefit analysis of all major regulations (those with an expected economic impact larger than $100 million). Directives published by the Office of Management and Budget (OMB) and the U.S. Department of Agriculture (USDA) clarify the general guidelines found in EO 12866. OMB's specific guidelines are found in Circular A-94 "Guidelines and Discount Rates for Benefit-Cost Analysis of Federal Programs." USDA guidelines are found in Appendix C of the Departmental Regulation on Regulatory Decisionmaking, DR 1512-1, "Guidelines for Preparing Risk Assessments and Preparing Cost-Benefit Analyses."

that potentially pose biological and toxicological risks. In the first section of the chapter I examine the origins and principal provisions of the SPS Agreement. In the following section I provide a brief review of the use of cost–benefit analysis in regulatory decision making, which figures prominently in the Executive Branch initiatives. This review sets the stage for an assessment of the role of economic criteria in the multilateral rules for SPS measures. A close legal reading of the SPS Agreement is beyond the scope of this chapter and the expertise of this author. Rather, this review is intended to flag a number of potential issues and questions that could arise as the United States seeks to manage trade-related health and environmental risks more efficiently. The hope is that the answers to these questions can provide the basis for the design of an SPS regulatory template that gains international standing. In the final section I present some brief concluding remarks about the potential role for economics in risk management policies. This discussion notes that at this stage—in advance of extensive SPS jurisprudence and before many WTO members have acquired an understanding of their new international rights and obligations—the development of principles for efficient regulatory decision making could make a substantial contribution to the international trading system.

THE SPS AGREEMENT: ORIGIN AND PRINCIPAL PROVISIONS

Origin of the SPS Agreement

Prior to the conclusion of the 1986–1993 Uruguay Round of trade negotiations, multilateral disciplines on the use of SPS measures were found in the original General Agreement on Tariffs and Trade (GATT) articles (primarily Article XX, *General Exceptions*) and the 1979 Tokyo Round Agreement on Technical Barriers to Trade (which was a plurilateral agreement known as the Standards Code). Although these legal instruments stipulated that measures could not be "applied in a manner which would constitute ...a disguised restriction on international trade" or "create unnecessary obstacles to trade," the consensus view that emerged in the decade following the Tokyo Round was that multilateral rules had failed to stem disruptions of trade in agricultural products caused by proliferating technical restrictions (Roberts, 1998). Not one SPS measure was successfully challenged before a GATT dispute settlement panel after the Tokyo Round, and several prominent disagreements over SPS measures in the 1980s (most notably the U.S.–European Union beef dispute over hormone-treated beef) remained unresolved (Stanton, 1997). Meanwhile, the commitment to negotiate an Agriculture Agreement during the Uruguay Round, which would discipline the use of agricultural nontariff barriers for the first time, heightened concerns that governments would resort to regulatory compensation in the form of SPS barriers to appease domestic producers in this politically sensitive sector (Josling et al., 1996).

Consequently, the Punta del Este Ministerial Declaration, which launched the Uruguay Round in 1986, stated that one objective of the negotiations would be to create disciplines that would minimize the "adverse effects that sanitary and phytosanitary regulations and barriers can have on trade in agriculture." Initial negotiations targeted perceived defects in the Standards Code, which had impeded resolution of some SPS disputes.[3] But despite progress on closing some loopholes in early drafts of new Technical Barriers to Trade (TBT) Agreement, support for the negotiation of a separate SPS Agreement emerged during the negotiations. Negotiators concluded that multilateral rules for adoption of risk-reducing trade measures, which routinely violate GATT Most-Favored-Nation (MFN) and National Treatment principles[4] could not be conveniently incorporated into the new TBT Agreement. In 1988, a separate working party was created to draft an SPS Agreement. The working party, which included representatives from agricultural and trade ministries, as well as regulatory agencies, produced an agreement that established new substantive and procedural disciplines for SPS measures. The substantive requirements suggest a normative basis for SPS measures, whereas the procedural obligations facilitate decentralized policing of such measures.

Principal Provisions of the SPS Agreement

The SPS Agreement consists of a preamble that states the objectives of the agreement in broad terms, 14 articles that stipulate both procedural and substantive disciplines, and 3 annexes that set forth definitions and elaborate on procedural requirements (GATT, 1994). Articles 2 through 6, together with the definitions of *SPS measure, risk assessment,* and *appropriate level of protection* found in Annex A of the agreement, provide a basis for understanding what the principal multilateral rules for SPS measures are in the post-Uruguay-Round era.

The disciplines apply to regulations defined as SPS measures by the agreement: those measures that protect human, animal, or plant life and health within the territory of the member from risks related to diseases, pests, and disease-carrying or disease-causing organisms, as well as additives, contaminants, toxins, or disease-causing organisms in food, beverages, or feedstuffs. Two important points emerge from the definition. First, SPS

---

[3]For example, one loophole had been created by the Standards Code's definition of a measure that would be subject to the disciplines in the agreement. It defined a technical regulation as "A specification contained in a document which lays down characteristics of a product such as levels of quality, performance, safety or dimensions."—which omitted explicit reference to production and processing methods. This omission provided the legal rationale for the European Community when it blocked the U.S. request for a technical experts group to review the scientific basis of the European Community ban on hormone-treated beef in 1987.

[4]SPS measures mitigate risks that may vary by the source and destination of a traded good. SPS measures are therefore more likely than other technical measures to violate the MFN principle found in Article I of the GATT (which stipulates that concessions offered to one trading partner must be offered to all) or the National Treatment principle codified in GATT Article III (which holds that imported products be "accorded treatment no less favorable than that accorded to like products of national origin").

measures are defined with respect to the regulatory goal of a measure rather than the policy instrument itself (in contrast to other WTO disciplines for specific policy instruments such as tariffs). Thus, SPS measures span policy instruments of differing degrees of trade restrictiveness, from complete bans to information remedies (e.g., labels that list known allergens). Second, "plants and animals" in the definition include natural flora and fauna, and therefore SPS disciplines also apply to measures intended to protect unowned or commonly owned environmental assets. The agreement thus disciplines measures that protect both market and nonmarket goods.

The cornerstone of the SPS Agreement is Article 5 (*Assessment of Risk and the Determination of the Appropriate Level of SPS Protection*). Article 5.1 contains the explicit requirement to base decisions about SPS measures on a risk assessment, which is defined as the evaluation of the likelihood and biological and economic consequences of identified hazards under different risk management protocols.[5] Article 5.3 stipulates that countries are to consider direct risk-related costs (e.g., potential production or sales losses, administrative expenses, potential eradication costs) in risk management policies for plant and animal health. Members are also obligated to take into account the relative cost effectiveness of alternative approaches to limiting risks.[6]

Four other articles comprise the remaining substantive disciplines in the SPS Agreement. Article 2 (*Basic Rights and Obligations*) stipulates that measures must be "based on scientific principles," "not maintained without sufficient scientific evidence," and "applied only to the extent necessary." It also states that members must ensure that their measures do not arbitrarily or unjustifiably discriminate between members where identical or similar conditions prevail, which includes between their own territory and that of other members.[7] Thus, if the commodity risk is thought to be the same for imports from country *X* and *Y*, the language of the agreement suggests that the importing country should adopt the same import measure for both countries. Article 6 (*Adaptation to Regional Conditions, Including Pest- or Disease-Free Areas and Areas of Low Pest or Disease Prevalence*) codifies the same modified MFN and National Treatment principles for subnational units that are free from diseases and pests or where the prevalence of diseases and pests are low. Article 3 (*Harmonization*) stipulates that, although members can adopt a measure to

---

[5]Factors that should be taken into account in a risk assessment—available scientific evidence; relevant processes and production methods; relevant inspection, sampling, and testing methods; prevalence of specific diseases or pests; existence of pest- or disease-free areas; relevant ecological and environmental conditions; and quarantine or other treatment—are specified in Article 5.2.

[6]Cost-effective analysis is a special case of cost-benefit analysis in which the regulatory goals are fixed (generally in terms of physical outcomes) and the analysis is an attempt to identify the least-cost means of achieving them.

[7]These latter disciplines are variants of the GATT MFN and National Treatment principles, modified as suitable to the circumstances posed by the biological and toxicological risks at issue.

provide a higher level of health or environmental protection than that provided by an existing international standard, scientific evidence must support that claim, and Article 4 (*Equivalence*) states that an importing country must accept a foreign measure that differs from its own as equivalent if the foreign measure provides the same level of health or environmental protection.

The phrase "appropriate level of protection" is threaded throughout the agreement, from the preamble to the annexes, clearly reinforcing the fact that the risk assessment paradigm was the point of reference for the SPS Agreement negotiators. For example, the agreement states that members may adopt measures that do not conform to international standards if these standards do not provide the level of SPS protection that a member determines to be appropriate (Article 3); that members shall accept the SPS measures of other members as equivalent if the exporting country objectively demonstrates that its measures achieve the importing country's appropriate level of SPS protection (Article 4); and that members shall avoid arbitrary or unjustifiable distinctions in appropriate levels of protection if such distinctions result in discrimination or a disguised restriction on international trade (Article 5). A member's appropriate level of SPS protection is tautologically defined in the agreement as the level of protection deemed appropriate by the member establishing SPS measures to protect human, animal, or plant life or health. An explanatory note states that many members otherwise refer to this concept as the "acceptable level of risk."

There are some implications of adoption of a risk assessment paradigm. In this paradigm, analysts typically identify measures that are determined to mitigate risks to acceptably small levels. The risk management decision—in this case, the choice of an import measure or protocol—is restricted to consideration of the set of measures identified by analysts as achieving the risk target. The determination of the acceptable level of risk or appropriate level of protection encourages a myopic focus on only the direct risk-related costs of import protocols. The potential benefits of different regulatory options are only intermittently factored into decisions that regulators view as "unarbitrary and justifiable" distinctions in the appropriate level of protection. For example, it is not uncommon for regulators to accept imports of live breeding stock while rejecting meat because of "industry needs." The role of economics in the risk assessment paradigm is relegated primarily to the calculation of the quantity of imports to help risk assessors with their job of calculating the likelihood and consequences of disease or pest introduction, rather than to provide an explicit accounting of the costs and benefits of a policy's effects on producers, consumers, taxpayers, and industries that use the regulated product as an input.

COST–BENEFIT ANALYSIS OF SPS REGULATIONS

U.S. Executive Branch directives related to improvement in the content and process for regulations over the past 25 years reflect policymakers' increasing interest in the use of cost–benefit analysis (CBA) as a tool for regulatory assessment. The intellectual foundations of CBA are found in welfare economics, which provides a theoretical framework within which policies can be ranked on the basis of how much they improve social well-being. Social welfare is the yardstick used by economists to provide a single metric that captures the relevant features of well-being that might be affected by a policy. CBA can be described simply as a study to determine what effect proposed alternative policies would have on the value of this social welfare metric.[8] Although acknowledging the empirical challenges associated with CBA, it is nonetheless advocated by many as a means for conveying some normative information to decisionmakers.[9] Its principal merits include transparency, a consistent framework for data collection and characterization of information gaps, and the ability to aggregate dissimilar effects into one measure (Kopp et al., 1997).

The metric employed in CBA is a monetary measure of the aggregate change in individual well-being resulting from a policy decision. In the economic paradigm, individual welfare is assumed to depend on the satisfaction of individual preferences, and monetary measures of welfare change are derived from the measurement of how much individuals are willing to pay or to be compensated to live in a world with the policy in force. Within this paradigm, a policy that improved social welfare as indicated by the metric would be preferred to a policy that would reduce welfare, and a policy that would increase welfare more would be preferred to a policy that would increase welfare less. Because a policy can (and indeed usually does) make some individuals better off while making other individuals worse off, the metric employed in CBA is net social welfare. The use of net social welfare to guide policy choices rests on the compensation principle—that a policy is preferred to the status quo if all those who benefit from the policy could *in principle* compensate those who lose, and still be better off than before the policy went into effect.

Net social welfare, or net benefits, produced by alternative SPS measures can be described most easily in the context of a single-commodity, partial equilibrium model to evaluate a proposed change in a plant or animal health measure. In this simple framework, SPS policy evaluation would entail the calculation of changes in the welfare of producers and consumers of the regulated product.[10] For example, quarantine policy prohibits agricultural products from foreign sources unless the U.S. Department of Agriculture

---

[8]I am grateful to one reviewer for pointing out that multiattribute decision analysis tools, which likewise reflect an economic perspective on risk management, can be used when aggregation of dissimilar effects into one measure is impossible or deemed to be inappropriate. See, for example, Kasperson (1992) and Raiffa (1968).

[9]There are objections to the use of CBA on both philosophical and pragmatic grounds. The pragmatic concerns are taken up in the following section. The primary focus of this study is *whether* and *how* economic criteria can be used in the development of decision rules without violating the letter or spirit of the SPS Agreement, not the question of should these criteria be used.

[10]This discussion assumes that the only goods protected by the border measure are market goods such as herds and flocks. The issue of nonmarket goods, such as wildlife, is taken up in the next section.

(USDA) has specifically determined if and under what conditions a product may enter the United States. In this case, CBA-based decisions would require evaluation of whether the benefits of lower-priced imports (to consumers) would outweigh the potential costs (to producers) associated with these same imports. Producer losses would stem from two sources in this open-economy framework: lower product prices and the expected value of losses resulting from exotic pests or diseases.

The net benefits of a proposed measure would be calculated from changes in producer and consumer surplus. Producer surplus is defined as producers' revenue beyond variable costs which provides a measure of returns to fixed investment. Consumer surplus is a measure of consumers' willingness to pay for a product beyond its actual price. Producer and consumer surplus can be seen in the context of the partial equilibrium model characterized in Figure 2-1. If the country did not import the product from any source prior to the import request received by the USDA, the price of the product in the domestic market (P_D) would be determined by the intersection of the domestic supply and demand curves (respectively denoted S_1 and D_1). In autarky (when there is no trade), the quantity demanded and supplied in the domestic market is Q_1. In this scenario, the area bounded by the demand curve and the price line is consumer surplus (areas $A + B$), and producer surplus is the area bounded by the supply curve and $P_D (C + D + G + H + I + J)$.

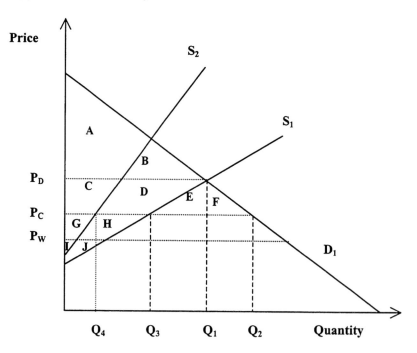

FIGURE 2-1. A Partial Equilibrium Model of the Welfare Effects of Alternative Import Protocols

Assume now that the USDA approves the import request, which allows imports of the product under a specified import protocol. These imports lower prices in the domestic market to P_C, which is equal to the world price P_W, plus compliance costs associated with the protocol.[11] (If the import request came from a country that produced the product at a higher compliance-cost inclusive price than the U.S. price, imports would not occur.) At P_C, the quantity demanded by domestic consumers increases to Q_2 while the quantity supplied domestically shrinks to Q_3. (Imports equal to $Q_2 - Q_3$ make up the difference.) If scientists and regulatory officials judged that the probability of importing a disease along with the product was essentially zero—for example, if a disease had never been known to exist in the exporting country—producers would lose the area $C + D$, while consumers would gain the area $C + D + E + F$. The net benefits of this decision would be $E + F$. This "zero-risk" scenario is identical to the standard trade liberalization scenario wherein a country decides to eliminate a tariff.

However, evaluation of a change in an SPS measure to allow imports differs from the evaluation of removing a tariff if there is some probability, however small, that a disease will be imported along with the product. In addition to the producer and consumer surplus changes that result from a decrease in price in the domestic market from P_D to P_C, potential production losses from a disease must be evaluated as well. If pests raise production costs and lower yields with certainty, domestic supply will shift from S_1 to S_2, leading to a decline in domestic production to Q_4 in Figure 2-1. Assuming trade is not embargoed, imports increase to $Q_2 - Q_4$. In this scenario, producers would lose $C + D$ (the trade effect) as well as $H + J$ (the disease effect), while consumers would gain $C + D + E + F$ as before. In this simplified example, consumers are always better off, producers are always worse off, and the net benefit of this regulatory option is $(E + F)$ minus $(H + J)$, a difference that can be positive or negative. On a probabilistic basis, the expected domestic supply function will lie between S_1 and S_2 with its location depending on the assumed level of disease risk under the protocol. $E + F$ is likely to be bigger than $H + J$ in instances where there is a large price difference between the domestic and world price of a product, and the probability of introducing a relatively innocuous disease is negligible. $H + J$ will be bigger than $E + F$ in instances where there is little difference between the world and domestic price of a product, and the probability of importing an especially virulent disease is high. It is important to note that the choice among different risk mitigation alternatives will simultaneously determine the location

---

[11]Two assumptions underpin the analysis presented in this discussion. First, it is assumed that the importing country is small in terms of the world market for the product, so its trade volume will not affect the world price. Hence, the excess supply curve faced by the importing country is perfectly elastic at P_W. Second, Figure 2-1 reflects the assumption that the same regulation applies to all exporters and raises the price of the imported good by a fixed per unit amount, so that the compliance-cost inclusive price for the imported good can be characterized as P_C.

of S_2 and P_C, which complicates the evaluation of changes in producer and consumer surplus.

A decision rule based solely on CBA would imply that the decisionmaker would choose the import measure that would maximize the difference between $E + F$ and $H + J$ in this highly stylized example. Actual empirical estimates of the areas pictured in Figure 2-1 under different import protocols pose substantial challenges. The analyst must have a means for assessing the probability of introduction of a disease (a likelihood model) together with a means of assessing different disease outcomes (an epidemiological model) as inputs into the economic model to estimate (1) the expected value and standard errors of changes in producer surplus stemming from disease-related production and sales losses, (2) producer surplus losses resulting from lower prices, and (3) consumer gains from lower prices. Model results could then provide one input into the calculation of costs not pictured in Figure 2-1, including the administrative costs of the import protocol (if not covered by user fees) and the expected value of government disease eradication expenditures.

Earlier research on the costs and benefits of SPS measures typically overlooked the social welfare losses caused by restricting imports. For example, Aulaqi and Sundquist's CBA of the U.S. ban on imports from countries with foot-and-mouth disease (FMD) estimated the disease-related economic losses (a measure of the benefit of keeping the ban in place) and compared it to the public expenditures necessary to enforce the ban (a measure of the cost of the ban) (Paarlberg and Lee, 1998). Yet restrictive trade measures such as import bans impose welfare costs on society through higher prices, information that is omitted in studies that fail to account for the wider market effects of SPS policies.

How might a regulatory decision be altered by a full accounting of the costs and benefits resulting from a change in an import measure? By way of example, one could consider a recent study that calculated the costs and benefits of the USDA's 1996 rule which replaced an 80-year-old ban on imports of Mexican Hass avocados with a geographical and seasonal protocol (Orden and Romano, 1996). The new import protocol allows Mexican avocados to be exported to the Northeastern United States during winter months. In a long run model of the new measure, the estimates of net welfare increases ranged from $2.5 million to $55.7 million under different pest infestation scenarios.[12] Had the ban been completely rather than partially rescinded, the estimates of net social benefits would have ranged from $32.4 million to $13.9 million under the same set of infestation assumptions. The primary difference between the alternatives is that, by completely rescinding the ban, consumers throughout the United States

---

[12]Pest infestation assumptions range from a $1.35E^{-06}$ probability of a 20 percent increase in marginal costs with no reduction in yield to a certain occurrence of a 60 percent increase in marginal costs plus a 20 percent reduction in yields. Estimates of the probabilities of pest introduction were drawn from risk assessments introduced into the public record by the USDA and the domestic industry (Firko, 1995; Nyrop, 1995). The impacts of a pest introduction on production costs were drawn from an unpublished USDA study (Evangelou et al., 1993).

would be able to purchase avocados at significantly lower prices year round, resulting in substantial increases in the benefits of this option relative to the alternative geographical and seasonal restrictions. If the decision rule were based solely on a CBA of the alternatives, the avocado ban would have been lifted altogether.[13]

Another recent study of Australia's total ban on banana imports reinforces the same point—that in some instances the consumer (as well as processor and retailer) gains from removing a ban can far outweigh any loss to growers, even if diseases were to wipe out the industry (James and Anderson, 1998). In this sense, "bad" phytosanitary policy can be "good" economic policy. The authors note that compensating producers out of consolidated revenue might be an affordable way to reduce opposition to changes in quarantine policies in these cases.

Although a single-commodity, partial equilibrium framework captures the direct costs and benefits for producers and consumers of a product (and is often the most feasible tool for analyzing specific regulatory proposals), it is important to recognize that there are always indirect economy-wide effects, and these can be larger than direct effects. Most importantly, any SPS restriction that increases the price of an imported good is, in effect, a tax on all exports. Raising the price of a tradable good bids resources away from other industries. Eliminating unnecessary SPS measures or improving the design of necessary measures to allow more imports allows trading nations to reallocate productive resources toward economic activities in which they have a comparative advantage. Modeling these global efficiency gains requires extending the analysis to include inputs, multiple commodities, and multiple countries in a general equilibrium framework.

IS THE SPS AGREEMENT CONGRUENT WITH EXECUTIVE BRANCH GUIDELINES?

From the previous discussion, what can be said about the apparent congruence or incongruence of the SPS agreement with recent U.S. Executive Branch directives? Examination of this issue is made somewhat difficult by the fact, that in aiming to avoid being overly prescriptive, both the SPS agreement and the CBA guidelines provide latitude for alternative interpretations.[14] The

---

[13]This statement assumes that all individuals are equally weighted by the decisionmaker. This is not an inherent requirement of CBA, but rather the standard default assumption in view of the fact that U.S. Executive Branch directives and guidelines typically do not explicitly specify weights to reflect distributional concerns.

[14]Some reasonable interpretations of the individual provisions of the SPS Agreement can lead one to conclude that the agreement is internally inconsistent. In the words of the Appellate Body, some parts of the agreement are "not a model of clarity in drafting and communication" (WTO, 1998, p. 66).

following discussion should be understood as only an attempt to flag potentially important issues for further discussion as the United States considers how best to incorporate CBA into its SPS decisions.

Disciplines that pertain to costs, benefits, and distributional effects of SPS measures are considered in turn. It is argued here that the language of the SPS Agreement, a product of the risk assessment paradigm, is clearest with respect to the risk-related costs associated with regulatory actions. How benefits of alternative regulatory options may factor into decisions is ambiguous, an ambiguity that slowed progress in fulfilling the mandate stipulated in the SPS Agreement to develop guidelines for measures to achieve the objective of providing a consistent "appropriate level of protection." Although distributional effects are not explicitly addressed in the agreement, it is not difficult to see how it does and does not circumscribe a regulator's ability to take distributional factors into account in risk management decisions.

Costs

Article 5.3 is reprised here to facilitate comparison of costs that are recognized by the SPS Agreement and costs as they are routinely calculated in a CBA. The article states

> In assessing the risk to animal or plant life or health and determining the measure to be applied for achieving the appropriate level of sanitary or phytosanitary protection from such risk, Members shall take into account as relevant factors: the potential damage in terms of loss of production or sales in the event of the entry, establishment or spread of a cost or disease; the costs of control or eradication in the territory of the importing Member; and the relative cost-effectiveness of alternative approaches to limiting risks.

The cost components of a standard CBA were identified in the previous section as (1) the expected value of changes in producer surplus stemming from disease-related production and sales losses, (2) producer surplus losses resulting from lower prices, (3) government expenditures (if not covered by user fees) in administering the import protocol, and (4) the expected value of public-financed disease eradication expenditures.

With respect to the first item, the agreement seems to neither endorse or prohibit the translation of the expected value of revenue (production and sales) losses estimated during the course of a risk assessment into the expected value of producer surplus losses in a CBA framework. The agreement explicitly allows (and indeed uses the verb "shall" which indicates a legal obligation) the third and fourth components to factor into risk management decisions. Consideration of the second component would likely be seen to be in violation of the spirit of the SPS Agreement. The agreement is predicated on the idea that countries should not factor lower product prices resulting from imports into a

SPS decision. Such producer surplus costs would be regarded as costs related to commercial activity, unrelated to health or environmental protection.[15]

So far, the discussion of regulatory measures throughout this paper has centered on SPS measures that protect market goods. However, SPS measures include border measures that protect native flora and fauna as well. In many circumstances one measure will simultaneously protect herds as well as wildlife, because diseases can affect both—FMD, for example, affects both cattle and deer. For a complete accounting of the potential societal costs of a regulatory decision to allow imports that could introduce FMD into the importing country, one would ideally have to also estimate the value that the society placed on deer. Formally, an economist could estimate the recreational or nonuse value of an environmental good by means of stated preference techniques (e.g., contingent valuation methods) or revealed preference techniques (e.g., travel cost methods). Sometimes, however, the value of an unowned or commonly owned resource is (implicitly or explicitly) set by legislation (e.g., the Endangered Species Act). The SPS agreement says nothing about how protection of nonmarket environmental assets can or should be factored into risk management decisions.

Benefits

It is interesting to note that the word *benefits* (in reference to trade or anything else) does not appear in the agreement.[16] One's interpretation of this omission is likely to depend on how one views the GATT and the WTO. One view is that the recognition of the benefits of trade liberalization is so universal that it merits no further emphasis. A less magnanimous view is that the trading system built around the GATT over the past 50 years is the product of enlightened mercantilism, rather than the ideology of free trade.[17] Statements found in the agreement such as "sanitary measures shall not be applied in a manner which would constitute a disguised restriction on trade" (Article 2.3) or "Members shall insure that such (SPS) measures are not more trade-restrictive than required to achieve their appropriate level of sanitary or phytosanitary protection" (Article 5.6) are broad enough to accommodate both views.

---

[15]For example, Article 2.2 reads "Members shall ensure that any sanitary or phytosanitary measure is applied only to the extent necessary to protect human, animal or plant life or health, is based on principles and is not maintained sufficient scientific evidence..."

[16]Recall that from the point of view of a change in quarantine policy in response to an import request, benefits are the gains in consumer surplus that result from lower prices in the domestic market (resulting from the entry of lower-priced imports) plus the elimination of dead weight losses from the trade barrier.

[17]Krugman (1991) observes that GATT-think (i.e., exports are good; imports are bad; other things equal, an equal increase in imports and exports is good) sees trade policy as a prisoners' dilemma: Individually, countries have an incentive to be protectionist, yet collectively they benefit from free trade.

In view of the lack of explicit disciplines on if and how the gains from trade can factor into SPS regulatory decisions, what conclusions can regulators draw? An example can best illustrate the agreement's ambiguity about the standing that benefits have in SPS regulatory decisions. Consider an example in which the USDA decides to allow imports of beef, but not poultry, because, although the expected value of disease-related losses are the same for the two products, the benefits to consumers of importing beef outweigh those costs whereas the benefits of importing poultry do not (i.e., relative to foreign competitors, the United States is a more efficient producer of poultry than of beef). Although the choice to allow only imports of beef might be efficient regulatory policy from a CBA perspective, some in the WTO community (and most certainly the country whose import request for poultry was turned down) would view these choices as evidence of "arbitrary and unjustifiable distinctions" in the levels of protection that had resulted in discrimination or a disguised restriction on trade, in violation of Article 5.5.

The omission of explicit disciplines on the consideration of benefits in the SPS agreement can perhaps be defended on the grounds that the agreement is an international trade treaty, not a regulatory blueprint. The purpose of the legal obligations in the agreement is to limit the use of the putative scientific claims for protectionist purposes, not to establish templates for risk management decisions. In short, the trade disciplines were not intended to be good practice standards.

But one might have hoped to see good practice standards emerge from the SPS Committee's efforts to develop guidance for Members to further the practical implementation of Article 5.5, as mandated by the Agreement. It has been clear in Committee debates, however, that many WTO Members hold the view that consumer/processor gains from trade fall in the same category as producer losses that result from decreases in domestic prices brought about by imports—as *commercial* considerations that might be appropriately factored into a country's choice of its single "appropriate level of protection," but which should not be used as decision criteria for individual risk mitigation measures. This view has stemmed from both philosophical objections as well as pragmatic concerns that CBA-based import protocols would complicate the effective decentralized policing of SPS measures by WTO Members. The Committee is scheduled to adopt a set of non-binding guidelines in June 2000, after struggling with its mandate over the past five years. It appears as though WTO endorsement of the use of economic criteria in risk management decisions will, for the present, remain restricted to the consideration of risk-related costs.

Distributional Issues

It would appear that many animal and plant health import measures in both developed and developing countries reflect the fact that regulators have historically placed greater (implicit) weight on producer rather than consumer welfare. Over time, the net costs of some extremely conservative import protocols have likely risen as technological advances in transport have dramatically reduced economic distances. Executive Branch initiatives that include the requirement for the calculation of net social welfare for major regulatory actions can be viewed in the context of quarantine measures as an effort to prompt regulators to reexamine these implicit weights.

However, the directives reflecting the current mainstream view also state that, although net social benefits should be an important factor in regulatory decisions, it need not be the only one. For example, the USDA's Departmental Regulation on Regulatory Decisionmaking (DR-1512-1) encourages regulators to consider a broad range of qualitative factors such as equity, quality of life, and distributions of benefits and costs (USDA, 1996). These guidelines reflect societal concerns over a strictly utilitarian approach to policy making (e.g., adopting a policy that results in substantial benefits for the wealthy while impoverishing the poor).

Nothing in the SPS agreement precludes a member from maintaining extremely conservative import protocols to protect animal and plant health that favors producers over consumers, as long as there is some scientific basis for the measure. In fact, the agreement can be read as explicitly protecting the right of members to do so in the language regarding the choice of "appropriate levels of protection." According to the U.S. Statement of Administrative Action (SAA) to Congress, the agreement "explicitly affirms the rights of each government to choose its levels of protection including a 'zero risk' level if it so chooses" (President of the United States, 1994, p. 745). The SAA also notes that "In the end, the choice of the appropriate level of protection is a societal value judgment." Thus the agreement places no constraints on the USDA's choice of weights for producer and consumer welfare in a CBA framework as long as any variation in the weights does not appear to be connected to creating disguised restrictions on trade.

The use of other distributional effects as SPS decision criteria may be more limited by the agreement. For example, adopting more conservative protocols to provide a greater degree of sanitary protection for certain types of livestock because they are an important component of the income of poor farmers clearly runs counter to many basic SPS Agreement principles. However, potential conflict between the SPS Agreement and guidelines for consideration of distributional impacts in regulatory decisions is limited by the obvious fact that governments generally rely on other types of policies to remedy social ills.

The foregoing discussion highlights some issues that multidisciplinary teams will need to consider as U.S. agencies judge how best to incorporate CBA and other aspects of the Executive Branch directives into its regulatory decision-making process. One issue, the valuation of unowned resources, emerges within the risk assessment paradigm as well as the economic paradigm. Another, the issue of how policymakers should weigh the effects of a decision on producers and consumers, is implicit in the determination of the appropriate level of protection in the risk assessment paradigm and is explicit in the economic paradigm. Perhaps the most important issue is examination of the circumstances where using net benefits as a decision criterion (as recommended in the economic paradigm) might run afoul of provisions of the agreement that hinge on comparisons of disease-related costs (the end product envisioned in the risk assessment paradigm).

CONCLUSIONS

Further study of individual SPS measures will provide evidence about the degree to which the SPS disciplines contribute to good economic policy. To date, debate over SPS measures has generally focused on the roles of national sovereignty; consumer concerns; and risk assessment in policy formulation, primarily reflecting legal, political, and scientific perspectives on risk management. The economic perspective on SPS regulatory decision making— that regulatory decisions should be informed by an analysis of the costs and benefits of the proposed regulatory options—has not been prominent in these policy discussions. Perhaps the conclusion to be drawn from discussions in the SPS Committee and elsewhere is that it is inaccurate to portray the SPS Agreement as a binding constraint that prevents regulators from using economic efficiency criteria as guideposts in making risk mitigation decisions. It would appear that such criteria have not systematically factored into SPS regulatory decisions in many WTO member countries, either before or after the Uruguay Round. It may therefore be more accurate to view the SPS Agreement as a mirror rather than a yoke for current approaches to risk management. But despite differences between what economists would recommend and what the agreement might allow or proscribe, the risk assessment paradigm of the SPS Agreement has clearly reduced the degrees of freedom for the disingenuous use of SPS measures to restrict imports in response to narrow interest group pressures. Because the past four years have been witness to a number of unilateral and negotiated decisions to ease SPS trade restrictions, the principles and mechanisms established by the agreement are credited with being an important institutional innovation that has, in some instances, counterbalanced regulatory protectionism or prodded regulatory inertia.

It is from this perspective that others will monitor how the United States will allow an integrated assessment of the costs and benefits of mitigation alternatives to factor into decisions about regulations that govern if and how agricultural products gain accesses to U.S. markets. Therefore, the challenge is to develop a framework in which mitigation alternatives can be ranked on the basis of efficiency and distributional goals with sufficient transparency to permit judgment about compliance with specific SPS agreement obligations that WTO trading partners now expect. A truly integrated assessment will involve coordination of multiple disciplines: entomologists and epidemiologists; agronomists, ecologists, and veterinarians; and political scientists, philosophers, and economists. It is likely that differences in paradigms, unstated assumptions, and expected end products of analysis will make such collaboration difficult at first. But the new regulatory environment for SPS regulators in the United States demands that such challenges be met. One hope is that the intragovernmental discussions about the use of CBA as a normative tool for public decision making in the United States will help to clarify the international dialogue about criteria for the determination of levels of risk that are acceptable or appropriate.

REFERENCES

Evangelou, P., P. Kemere, and C. Miller. 1993. Potential economic impacts of an avocado weevil infestation in California. Unpublished paper, APHIS, U.S. Department of Agriculture, Washington, D.C.

Firko, M.J. 1995. Importation of Avocado Fruit (*Persea americana*) from Mexico: Supplemental Pest Risk Assessment. BATS, PPQ, APHIS, U.S. Department of Agriculture, Washington, D.C.

General Agreement on Tariffs and Trade (GATT). 1994. The Results of the Uruguay Round of Multilateral Trade Negotiations: The Legal Texts. Geneva: WTO.

James, S. and K. Anderson. 1998. On the need for more economic assessment of quarantine/SPS policies. Australian Journal of Agricultural and Resource Economics 42 (4):425–444.

Josling, T., S. Tangerman, and T.K. Warley. 1996. Agriculture in the GATT. London: Macmillan.

Kasperson, R. 1992. The Social Amplification of Risk: Progress in Developing an Integrative Framework. In Social Theories of Risk, S. Krimsky and D. Golding, eds. Westport, Connecticut: Praeger.

Kopp, R., A. Krupnick, and M. Toman. 1997. Cost-Benefit Analysis and Regulatory Reform: An Assessment of the Science and the Art. Discussion Paper 97-19. Washington, D.C: Resources for the Future.

Krugman, P. 1991. The move toward free trade zones. Economic Review, November/December:5–25.

Nyrop, J.P. 1995. A critique of the risk management analysis for importation of avocados from Mexico. Report prepared for the Florida Avocado and Lime Committee and presented at the public hearings on the avocado proposed rule, Washington D.C. and elsewhere, August.

Orden, D., and E. Romano. 1996. The avocado dispute and other technical barriers to agricultural trade under NAFTA. Paper presented at the conference NAFTA and Agriculture: Is the Experiment Working? San Antonio, Texas, November.

Organization for Economic Cooperation and Development (OECD). 1997. Regulatory reform in the agro-food sector, in OECD Report on Regulatory Reform, Sectoral Studies, Vol. 1. Paris: OECD.

Paarlberg, P. and J. Lee. 1998. Import restrictions in the presence of a health risk: an illustration using FMD. American Journal of Agricultural Economics 80(1):175–183.

President of the United States. 1994. Message from the President of the United States Transmitting the Uruguay Round Trade Agreements to the Second Session of the 103rd Congress, Texts of Agreements Implementing Bill, Statement of Administrative Action and Required Supporting Statements, House Document 103-316, Vol. 1:742–763.

Raiffa, H. 1968. Decision Analysis: Introductory Lectures on Choice Under Uncertainty. Reading, Mass: Addison-Wesley.

Roberts, D. 1998. Preliminary assessment of the effects of the WTO agreement on sanitary and phytosanitary trade regulations. Journal of International Economic Law 1(3):377–405.

Stanton, G. 1997. Implications of the WTO agreement on sanitary and phytosanitary measures. In Understanding Technical Barriers to Agricultural Trade, D. Orden and D. Roberts, eds. St. Paul, Minn: International Agriculture Trade Research Consortium.

USDA (U.S. Department of Agriculture). 1996. Guidelines for preparing risk assessments and preparing cost-benefit analyses, Appendix C, Departmental Regulation on Regulatory Decisionmaking, DR-1512-1. Available online at <http:/www.aphis.usda.gov/ppd/region/dr1512.html>.

WTO (World Trade Organization). 1998. EC Measures Concern Meat and Meat Products (Hormones), Complaint by the United States. Report of the Appellate Body, WT/DS48/AB/R. Geneva: WTO.

3

An Overview of Risk Assessment

JOHN D. STARK
Ecotoxicology Program, Department of Entomology,
Washington State University

The world we live in is a risky place. Every day the act of getting out of bed and facing the world exposes us to potential harm or even death. From the hole in the ozone layer, global warming, consumption of coffee and alcohol, the drive in the car to work, danger is around every corner—the message is loud and clear: We are at risk! We are deluged with reports from the media about the risks of many of the things we do, consume, or are exposed to every day. Not only are humans at risk, the very world we live in is at risk, and to top it off, we are to blame. Some risks are easy to quantify. The connection of obesity to diabetes, smoking to lung cancer, excessive consumption of alcohol to liver damage, and the time spent driving a car to getting in an accident are quite straightforward and easily quantifiable. However, what about the low levels of pesticides in our diets, exposure to radon in our homes, or exposure to electromagnetic fields? These risks are much harder to quantify. What are the risks associated with the introduction of an exotic species to a country or new geographical region that does not already have this species? How does the introduction of exotic species impact humans, crops, domesticated animals, and the habitats that we wish to protect?

The risks of some behaviors or chemicals to human health are certainly real, but risk has become for some a new-age religion. For example, survivalist groups had formed because they were convinced that the world as we know it was coming to an end due to the Y2K computer bug.

Before we can discuss risk assessment, we must have a definition of risk. At its simplest, risk is the probability that harm will occur from a specific act. This definition covers a lot of ground. One can imagine just about anything having a risk associated with it even if it is slight.

RISK AND TRADE BARRIERS

The process of living exposes all organisms to various risks. For humans some of these risks are self-imposed such as smoking tobacco, which increases the risk of developing lung cancer. However, the risks that are important to trade differ somewhat from the examples mentioned above. Sanitary and phytosanitary (SPS) procedures have been established by many countries to protect their agricultural economy and natural environment (Gray et al., 1998). The goal of these procedures is to limit the entry of foreign pests and diseases in their respective countries.

Risks associated with SPS measures are broken down into three categories:

(1) Direct food risks—additives, contaminants, toxins, or disease-causing organisms in food. Some examples are hormones and antibiotics in beef, pesticide residues in crops, aflatoxin in grains, *Escherichia coli*, *Salmonella*, *Listeria*, and botulism toxin in various foods, and food additives (colorings, flavor agents, etc.).

(2) Introduction of exotic organisms—plant- or animal-carried diseases, pests, diseases, or disease-causing organisms. Some examples are various insect pests that are introduced in produce such as the Mediterranean fruit fly, species that are introduced in something other than a commodity such as the Asian long-horn beetle, infectious agents such as prions that cause mad cow disease, and weed pests like purple loose strife.

(3) Damage caused by exotic organisms—by entry, establishment, or spread of pests. Here the concern is the actual damage that may be inflicted on agricultural industries.

WHAT IS RISK ASSESSMENT?

The National Academy of Sciences defines risk assessment as "the determination of the probability that an adverse effect will result from a defined exposure" (NRC, 1983). Contrary to popular belief, risk assessment is not a science but rather a combination of science and expert judgment. Scientific data are used to develop an assessment of risk, but the risk assessor does not often have extensive data and has to make a judgment call. Furthermore, the type of data available are not uniform for each risk assessment. For some chemicals, for example, a complete toxicological profile will be available whereas for others the data may be much more limited.

The process of risk assessment varies from agency to agency within the United States depending on the type of risk being evaluated: pesticide residues in food, hazardous waste sites, introduction of exotic species, etc. However, the same basic principles are usually followed and consist of four steps: (1) hazard identification, (2) dose-response assessment, (3) exposure assessment, and (4) risk characterization.

Hazard Identification

The first step in risk assessment is to determine whether the agent in question is hazardous. If the agent being evaluated is a chemical, then basic information about its toxicity is required. If the agent is an organism, then basic information about its biology and life history are required. To put this into perspective, at least for chemicals, the toxicity of several common chemicals to rats or mice is listed in Table 3-1.

Dose-Response Assessment

Here a characterization of the relationship between the dose or concentration and the incidence of adverse effects in exposed populations is developed. This is based solely on scientific data. Dose can be thought of as chemical concentration or the number of individuals of an exotic species that are introduced to a geographic area over time.

The most commonly used measure of toxic effect is the LD_{50}. The LD_{50} is a statistically derived measure of the dose-response relationship and is an estimate of the lethal dose that causes 50 percent mortality of a group of organisms being studied. Other dose-response measures are estimated in the same manner; for example, a dose that causes 50 percent reduction in offspring. If a large enough group of organisms is exposed to increasing concentrations of a poison, a sigmoid curve is obtained when the cumulative percent affected (dead) is plotted against a dose or concentration (Figure 3-1).

At low concentrations no effect is observed, but as the concentration increases, some of the organisms begin to respond. The highest concentration where no effect is observed is the no observable effect concentration or level. The threshold is the lowest dose that elicits a response or the lowest observable effect concentration. Eventually a dose is reached that kills all of the organisms being evaluated (maximum effect). It is at this point that an increase in dose can have no further effect. It is difficult to derive the LD_{50} or the slope of the dose-response line from a sigmoid curve. Also, data points along the dose-response line below the threshold and above the maximum effect do not provide data that can be used in the estimation of the LD_{50}. Methods have been developed to straighten the dose-response curve and estimate the LD_{50}. The first statistical approach for dose-response data was proposed by Trevan (1927), but many

TABLE 3-1. Acute LD_{50}[a] Values of Selected Common Chemicals

Chemical	Oral LD_{50} (rat or mouse) (mg chemical/kg body weight)
Botulism toxin	<0.001 mg/kg
Aflatoxin B_1	9 mg/kg
Sodium fluoride	180 mg/kg
Tylenol	338 mg/kg
Diazinon	350 mg/kg
Aspirin	1,500 mg/kg
Malathion	2,800 mg/kg
Table salt (sodium chloride)	3,750 mg/kg
Ethanol (alcohol)	10,600 mg/kg

[a]LD_{50} is the lethal dose that kills 50 percent of a population.

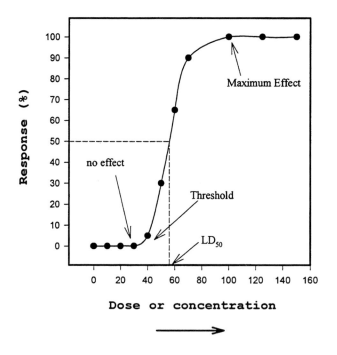

FIGURE 3-1. Dose–Response Relationship

modifications have been made over time (Finney, 1971). Acute toxicity data are usually analyzed by probit or logit analysis (Finney, 1971).

SELECTING TOXICOLOGICAL ENDPOINTS: WHAT DO WE EVALUATE?

Toxicological data used in risk assessments fall under two categories—acute and chronic exposure data. Acute data are generated after a single exposure to a chemical for a short time period. Chronic data are generated after repetitive exposure to concentrations over many days to a lifetime. Mortality is the endpoint of interest for many acute studies whereas life span, reproduction, weight gain, cancer, and birth defects are of interest in chronic studies.

To establish the amount of pesticide that can be ingested over a lifetime without causing illness, the lowest no observable effect level (lowest value for the endpoints studied—cancer, offspring, life span, etc.) is divided by a safety factor of 10–1,000 which results in the reference dose (RfD). The RfD is the dose or concentration below which daily aggregate exposure over a lifetime will not pose an appreciable risk to human health. What is an appreciable or acceptable risk? One new case of cancer in 1,000,000 people is considered acceptable. The reason that the safety factor varies has to do with the type of data available. Very little human toxicological data are available because we do not conduct toxicity studies with humans. What human data we do have usually come from accidents, suicides, or worker exposure. Therefore, extrapolation from animal data is often necessary. If human epidemiological data are available, then a safety factor of 10 might be used. If animal data are the only data available then a factor of 10 for the lack of human data is multiplied by 10 for animal data, resulting in a safety factor of 100 (10×10). The type of animal data available also reduces the safety factor. Chronic data result in a lower risk factor than acute data.

Exposure Assessment

A measure or estimate of the intensity, frequency, and duration of exposure agents is estimated in the exposure assessment part of risk assessment. All potential routes of exposure are considered in exposure assessment. For example, the likelihood of contact with the chemical through exposure to contaminated soil, water, air and/or food is evaluated. For chemicals this involves characterization of the exposure setting. The following questions are then asked:

- Where is the chemical likely to be found: water, soil, air, food?
- Are the organisms at risk aquatic or terrestrial or both?
- How will they be exposed? Through drinking, eating, breathing, dermal contact?

• How long will the exposure last?

A measure of the amount of chemical likely to be encountered in an environmental medium such as river water is estimated using data and mathematical models. For organisms such as disease organisms, the likelihood of exposure of a susceptible population over time is estimated.

Risk Characterization—What are the Consequences? Is there a Problem?

Risk characterization integrates the toxicity data and exposure assessment to arrive at probabilities of effects occurring.

There are several approaches to risk characterization. Perhaps the simplest is the quotient or ratio method (Barnthouse et al., 1986; Urban and Cook, 1986; Nabholz, 1991):

$$\text{Quotient} = \frac{\text{estimated environmental concentration}}{\text{toxicological endpoint concentration}}$$

The estimated concentration likely to be encountered is divided by the concentration estimated to cause a toxicological effect to arrive at the quotient. Quotients of 1 or greater imply a risk whereas quotients lower than 1 indicate less risk. For example, the LD_{50} for a pesticide to a fish species is 0.075 mg/l. The estimated environmental concentration is 0.1 mg/l. Using the quotient method, we find that 0.1/0.075 = 1.33, which means that this pesticide poses a risk to the fish species.

DETERMINISTIC RISK ASSESSMENT

The most commonly used methods of risk assessment today are deterministic and probabilistic risk assessment. A risk assessment based on a point estimate is called a deterministic risk assessment. Deterministic risk assessments are based on a single estimate of exposure (usually the worst-case scenario) and therefore do not provide information about variability and uncertainty that may be associated with a risk. The quotient method mentioned above is a type of deterministic risk assessment. However, deterministic risk assessment is often based on a tiered decision-making progress whereby a series of decisions are made based on the outcome of a previous result. As an example, imagine that a pesticide is registered for use on a hypothetical crop (crop A). The maximum allowable residue for the pesticide on crop A is 5µg pesticide/g of crop. An assumption is made that all of crop A is always sprayed with the maximum amount of pesticide allowed by the pesticide label, and thus the residue of pesticide present on the crop is also at a maximum. If the highest consumption of crop A is 10 g/kg body weight/day, then to arrive at the risk characterization, 5 µg pesticide/g of crop A × 10 g crop A/kg body weight/day = 50 µg pesticide/kg body weight/day. Exposure is then compared to the RfD. If the RfD is 75 µg pesticide/kg body weight/day, then there is no appreciable risk associated with this pesticide. Note that the assumptions made all err on the high side (are conservative). Refinements are made using the tiered approach that may result in less conservative (more realistic) estimates of risk.

PROBABILISTIC RISK ASSESSMENT

An assessment based on the probability of occurrence is called a probabilistic risk assessment. This method gives a measure of risk and the associated probabilities of their occurrence.

Using the pesticide and crop example above, data collected by the Food and Drug Administration (FDA) indicate that only 50 percent of the population consume crop A on any given day, and the amount consumed varies from 2, 4, 6, 8, and 10 g/kg body weight/day. Pesticide residues on the crop that are treated with the pesticide vary from 1, 2, 3, 4, and 5 µg pesticide/g of crop. Data from the Environmental Protection Agency indicate that only 25 percent of the crop is sprayed with the pesticide in question. The above data are run through a computer program (Monte Carlo simulation is an example of such probabilistic approaches) and the following exposures are generated:

- 78 percent of the population is not exposed to the pesticide,
- 1 percent is exposed to the highest exposure level (50 µg/kg body weight),
- 5 percent is exposed to 40 µg/kg body weight,
- 7 percent is exposed to 30 µg/kg body weight, and
- 9 percent is exposed to 10 µg/kg body weight.

The output is a distribution of risk values with a probability assigned to each estimated risk. Variability and uncertainty associated with the risk are part of the assessment (Hattis and Burmaster, 1994; Rai and Krewski, 1998). The general consensus among risk assessors is that probabilistic methods result in a risk assessment that is more realistic than a deterministic risk assessment.

PROTECTING HUMANS, PLANTS, AND WILDLIFE

Human Health—Pesticide Residues in Food

Obviously, protecting human health is the major concern for many risk assessors. Protection of human health, however, must be looked at in several different ways. The first and most obvious is direct protection, which is protection from disease-causing organisms and poisonings. Examples might be protection from diseases caused by organisms such as *E. coli*, *Salmonella*, *Listeria*, and parasites; and protection from toxins such as aflatoxin, botulin, pesticides, and hormones. To protect human health, consideration must also be given to clean water and air. Also, indirect protection involves protection of the human food supply and thus protection of crops and domestic animals.

As mentioned above, the reference dose is the amount of pesticide residue that can be ingested daily (daily allowable intake) by an average adult without an appreciable risk to human health.

How Safe is Our Food Supply in Terms of Pesticide Residues?

The percentage of foods that in 1997 contained pesticide residues is presented in Table 3-2 (FDA, 1998). Fruits had the highest percentage of residues and dairy products had the lowest. Interestingly, imported produce tended to have lower residues than commodities that originated in the United States. This is clearly the opposite of public perception.

As can be seen from the data presented in Table 3-2, very little of the agricultural commodities sold in the United States contain pesticide residues that are above the residue tolerance. However, pesticide RfD values are generated separately for each pesticide registration. The problem is that no one knows whether exposure to low levels of many pesticides (all at or below the RfD) can cause health problems. The reason for this is that the cost to do multiple exposure studies is prohibitive. To illustrate this problem, imagine that the cost to conduct a toxicological study for one chemical is $1,000. If we were to evaluate the toxicity of 10 chemicals including all possible combinations of these chemicals, the cost is 10 factorial (10!) × $1,000 or $3,628,800,000.

The Food Quality Protection Act (FQPA) was enacted in 1996. This law changed the way that the United States deals with pesticide residues in food. Prior to this law, residues for each pesticide were considered separately. The FQPA mandates that pesticides with like modes of action be lumped together. For example, residues of all organophosphate insecticides are added together to

TABLE 3-2. Pesticide Residues in Agricultural Commodities, 1997

| Commodity | Percent Commodity with Residues | | | |
| | Within Residue Tolerance | | Above Residue Tolerance | |
	Domestic	Imported	Domestic	Imported
Grains and grain products	40	13	0	1
Milk, dairy products, eggs	3	11	0	0
Fruits	55	38	1	1
Vegetables	28	35	2	1
Fish, shellfish, other aquatic products	32	6	0	0

Source: Adapted from FDA (1998).

come up with the total amount of exposure. This will result in the exceeding of tolerances for many classes of pesticide. The law then dictates that the total residue must be reduced. Pesticide uses on certain crops will almost certainly be reduced or eliminated all together in order to reduce total crop residues. Therefore, U.S. farmers stand to lose certain pesticides. However, pesticides that may be banned or their uses restricted in the United States will still be used in foreign countries. Some of these countries have developed accurate knowledge of the timing of sprays so that no residue is detected.

Plants, Domestic Animals, and Wildlife

When we think of plant protection, crops obviously come to mind, but we must also think of the risk posed to plant species that are not crops. Arthropod species and disease organisms that attack plants and weed species that compete with native species and other plants that we wish to protect are the major risk concerns.

Domestic animals are also susceptible to arthropod pests such as biting flies and disease organisms. Even weeds can be a big problem for our domestic animals. For example, some weeds are toxic to cattle.

No one would argue that protection of humans, domestic animals, crops, and wildlife from harm is important. However, when it comes to protecting wildlife, what should be protected? People in general like birds of prey, songbirds, sea mammals, salmon, and other fishes. But what about spiders, algae, and worms; are they not important as well? The most important species for ecosystem function may not be at the top of the food chain (the large predators). Our biases influence science and the funding of scientific research. Nowhere is this more evident than in environmental research. If two grants are submitted to a granting agency, the first dealing with determining the risks of pesticides to eagles and the second determining the risks of pesticides to soil-dwelling nematodes, guess which grant will get higher priority? Does this mean that eagles are more important to ecosystem function than nematodes? Not necessarily! In fact the nematode species in question may be more important than the eagle, but we assign a value to living things whether we realize it or not. Spiders, mites, algae, insects, and worms are just not high on our list of important species, yet the loss of these very organisms may be devastating to ecosystem function. Loss of eagles, a top predator, may not have much of an impact. Most people would agree, however, that the loss of eagles is unacceptable. We place a very high value on their presence in our world. Values are not the same for everyone. Wolves are a prime example. Ranchers hate them but conservation biologists and environmentalists love them.

Introduction of Exotic Species

Risk assessment was initially developed by the insurance industry in an attempt to determine life expectancies. A high degree of sophistication has been obtained in risk assessment of chemicals in food and the environment. However, risk assessment for biological hazards is much less developed than for chemical hazards (Powell, 1997).

This is due in part to the fact that the risk that introduced species pose is less subject to quantification. For many species we just do not know how well they will adapt to a new environment or whether they will change food sources or evolve. They may also interact with existing species in unpredictable ways.

One of the problems with conducting risk assessment of exotic species is determining how many organisms are necessary for establishment. Is one pregnant female of a potential insect pest enough to establish a population? Some pest species are parthenogenetic, that is, they are all females and produce clones of themselves. Thus, only one surviving individual may be enough to establish a population. For others species, many individuals may have to be introduced over time for establishment to occur. A great deal of knowledge about pest biology is therefore essential in developing pest risk assessment (Gray et al., 1998).

When it comes to trade of agricultural commodities, quarantines may be put in place that limit export from a particular geographical area (Gray et al., 1998). Postharvest disinfestation procedures, such as fumigation, may also be required (Stark, 1994) as well as inspections at points of export and import (Armstrong and Paull, 1994). Products may be banned if the risk is perceived to be very high or if there is no way to guarantee pest-free produce.

The following are some important questions that are asked by risk assessors about potential exotic pest species:

- Is a pest species present in an exporting country?
- Can the pest develop on hosts in the importing country?
- Can the pest species survive transport to the importing country?
- Are there quarantine treatments in place in the exporting country?
- How effective are the quarantine treatments?
- Can the pest exist in the climate of the importing country?

We should all be very concerned about the movement of species from one country to another because great economic and environmental damage can happen when exotic species arrive in a new geographic area. One of the greatest threats to wildlife is the introduction of exotic species because some of these species can outcompete native species and change the structure of communities of organisms.

Exotic species do not always enter a country directly on or in an agricultural commodity. An example of two species that have invaded the United States not through produce but related to commerce are the Asian long-horn beetle that entered the United States through shipping pallets originating in China, and the zebra mussel that entered the Great Lakes through the discharge of water ballast from ships originating in Europe. The zebra mussel, first discovered in 1988 in Lake St. Clair, is thought to have originated in the Caspian Sea. By 1990, zebra mussels were found in all the Great Lakes.

Zebra mussels are pests because they close off water supply pipes of various industries and power plants. They are also destroying native mussel populations through competition and directly by attaching to native mussels (Hebert et al., 1991; Hunter and Bailey, 1992; Nalepa, 1994; Schloesser and Nalepa, 1994; Ricciardi et al., 1995).

The Asian long-horn beetle was first discovered in the United States in Brooklyn, New York, in 1996 and has since been reported in Long Island, Chicago, and Bellingham, Washington. This species attacks hardwood trees with a preference for maples. Adult beetles chew holes in the bark and lay eggs. The larvae hatch and eat the bark, making tunnels as they grow. Mature larvae pupate and then the adult emerges from the tree by chewing through the bark. The Asian long-horn beetle kills the trees that it infests and thus is a very serious pest that could devastate many of our hardwood trees.

Transport of disease organisms is also a major issue in SPS measures. We only have to look at the recent outbreaks of *E. coli*-related foodborne illnesses in the United States to realize that food safety is a major concern worldwide. The recent outbreak of mad cow disease in the United Kingdom resulted in trade barriers being erected in other European Union nations. The presence of aflatoxins in grain and peanuts has also been a risk issue.

RISK ASSESSMENT OF GENETICALLY ENGINEERED ORGANISMS

An area that is already becoming a major trade issue is the importation of genetically modified organisms. One of the worries associated with the use of genetically modified organisms is the spread of genes from one species to another (Kareiva and Stark, 1994). Thus, a crop that is engineered to tolerate herbicides might transfer this gene through pollen to weed species resulting in weeds that are also resistant to a herbicide. The safety of food that has been genetically modified has also recently been called into question. In fact, trade of genetically modified agricultural commodities is presently being debated at the international level.

HOW CAN WE BE FOOLED? UNPROVABLE RISKS

As mentioned above, because of physical and financial limitations we do not know if exposure to multiple pesticide residues causes health problems. An argument could be made that food that contains several pesticide residues, even if they are at or below the RfD, could cause a health risk. It would be difficult to disprove this argument. A product that may appear harmless might actually cause a problem. And toxicologists know that exposure to low levels of poisons can actually result in increased vigor. How do we figure this into a risk assessment?

FUTURE PROBLEMS—SCIENTIFIC ARGUMENTS ABOUT RISK ASSESSMENT

Individual versus Population-Level Effects

One of the current debates in risk assessment has to do with the endpoints used to evaluate toxic effects. Toxicologists usually study the effects of chemicals in individuals. However, what happens at one level of organization (individuals) does not necessarily translate to another level (populations). The National Research Council (1981) has recommended that chemicals should be studied at the population-, community-, and ecosystem-level, yet few researchers have adopted approaches for the evaluation of chemical effects at levels of organization higher than the individual (Kareiva et al., 1996; Stark et al., 1997; EPA, 1998; Suter, G.W. II. 1999). One thing that may occur at the population level that cannot be accounted for by examining individual mortality and reproduction is "population compensation." For example, when individuals are removed from the population after exposure to a chemical, survivors have more resources available and may reproduce at a greater rate; offspring may also be larger and more vigorous. Thus, populations may be less susceptible than we predict based on studies with individuals. On the other hand, effects can be masked at the population level and loss of genetic diversity may occur.

Population Structure and Susceptibility

Susceptibility of a population may be greatly influenced by the structure of the population at the time of exposure. Toxicological studies are almost always conducted for one life stage or age. However, populations in nature often consist of a mixture of stages and ages. A recent study has indicated that the effect that pollutants have on populations is greatly influenced by the initial structure of the population at the time of exposure (Stark and Banken, 1999). These findings have implications for ecological risk and protection of wildlife. Susceptibility of populations in the wild may be greater or less than predicted depending on population structure.

CONCLUSIONS

Risk assessment is a valuable tool that combines science and expert judgment. Increasingly more sophisticated means of risk assessment have been developed, particularly in the areas of human health and the environment. However, risk assessment of exotic species is much less developed, and more work is needed in this area.

Trade disputes over food contaminants such as hormone and pesticide residues and exotic pest introductions have occurred in the past and may continue to be trade issues in the future, particularly in light of the FQPA. However, the issue that will probably dominate future trade disputes is genetically modified organisms.

REFERENCES

Armstrong, J., and R.E. Paull. 1994. Introduction. In Insect Pests and Fresh Horticultural Products: Treatments and Responses. R. E. Paull and J. Armstrong eds. Wallingford, U.K: C.A.B. International.

Barnthouse, L.W., G.W. Suter II, S.M. Bartell, J.J. Beauchampp, R.H. Gardner, E. Linder, R.V. O'Neill, and A.E. Rosen. 1986. User's Manual for Ecological Risk Assessment. ORNL Publication No. 2679. Oak Ridge, Tenn.: Oak Ridge National Laboratory.

EPA (U.S. Environmental Protection Agency). 1998. Guidelines for Ecological Risk Assessment. EPA/630/R-95/002F. Risk Assessment Forum. Washington, DC.: U.S. Environmental Protection Agency.

Finney, D.J. 1971. Probit Analysis, 2nd ed. Cambridge, U.K.: Cambridge University Press.

FDA (Food and Drug Administration). 1998. FDA Pesticide Program: Residue Monitoring. Available online at <http://vm.cfsan.fda.gov/~dms/pes97rep.html>.

Gray, G.M., J.C. Allen, D.E. Burmaster, S.H. Gage, J.K. Hammitt, S. Kaplan, R.L. Keeney, J.G. Morse, D.W. North, J.P. Nyrop, M. Small, A. Stahevitch, and R. Williams. 1998. Principles for conduct of pest risk analyses: report of an expert workshop. Risk Analysis 18:773–780.

Hattis, D. and D.E. Burmaster. 1994. Assessment of variability and uncertainty distributions for practical risk analysis. Risk Analysis 14:713–730.

Hebert, P.D.N., C.C. Wilson, M.H. Murdoch, and R. Lazar. 1991. Demography and ecological impacts of the invading mollusc *Dreissena polymorpha*. Canadian Journal of Zoology 69:405–409.

Hunter, R.D., and J.F. Bailey. 1992. *Dreissena polymorpha* (zebra mussel): colonization of soft substrata and some effects on unionid bivalves. The Nautilus 106(2):60–67.

Kareiva, P.K., and J.D. Stark. 1994. Environmental risks in agricultural biotechnology. Chemistry and Industry 2:52–55.

Kareiva, P., J.D. Stark, and U. Wennergren. 1996. Using demographic theory, community ecology, and spatial models to illuminate ecotoxicology. Pp.13–23 in Ecotoxicology: Ecological dimensions. L. Maltby and P. Grieg-Smith, eds. London: Chapman & Hall.

Nabholz, J.V. 1991. Environmental hazard and risk assessment under the United States Toxic Substances Control Act. Science of the Total Environment 109/110:649–665.

Nalepa, T.F. 1994. Decline of native unionid bivalves in Lake St. Clair after infestation by the zebra mussel, *Dreissena polymorpha*. Can. J. Fish. Aquat. Sci. 51:2227–2233.

NRC (National Research Council). 1981. Testing for Effects of Chemicals on Ecosystems. Committee to Review Methods for Ecotoxicology, Commission on Natural Resources. Washington, D.C.: National Academy Press.

NRC (National Research Council). 1983. Risk Assessment in the Federal Government: Managing the Process. Washington, D.C.: National Academy Press.

Powell, M. 1997. Science in Sanitary and Phytosanitary Dispute Resolution. Discussion Paper 97-50. Washington, D.C.: Resources for the Future.

Rai, S.N. and D. Krewski. 1998. Uncertainty and variability analysis in multiplicative risk models. Risk Analysis 18(1):37–45.

Ricciardi, A., F.G. Whoriskey, and J.B. Rasmussen. 1995. Predicting the intensity and impact of *Dreissena* infestation native unionid bivalves from *Dreissena* density. Can. J. Fish. Aquat. Sci. 52:1449–1461.

Schloesser, D.W., and T.F. Nalepa. 1994. Dramatic decline of native unionid bivalves in offshore waters of western Lake Erie after infestation by the zebra mussel, *Dreissena polymorpha*. Can. J. Fish. Aquat. Sci. 51:2234–2242.

Stark, J.D. 1994. Chemical fumigants. Pp. 69–84 in Insect Pests and Fresh Horticultural Products: Treatments and Responses, R.E. Paull, and J. Armstrong, eds, Wallingford, U.K.: C.A.B. International.

Stark, J.D., and J.A. O. Banken. 1999. Importance of population structure at the time of toxicant exposure. Ecotoxicology and Environmental Safety 42:282–287.

Stark, J.D., L. Tanigoshi, M. Bounfour, and A. Antonelli. 1997. Reproductive potential: It's influence on the susceptibility of a species to pesticides. Ecotoxicology and Environmental Safety 37:273–279.

Suter, G.W. II. 1999. A framework for assessment of ecological risks from multiple activities. Human and Ecological Risk Assessment 5(2):398–413.

Trevan, J.W. 1927. The error of determination of toxicity. Proc. Royal Soc. B 101:483–514.

Urban, D.J., and N. Cook. 1986. Ecological Risk Assessment. EPA 540/9-85-001. Office of Pesticide Programs. Washington, D.C.: U.S. Environmental Protection Agency.

4

Technological Risk and Cultures of Rationality

SHEILA JASANOFF

John F. Kennedy School of Government, Harvard University

The latter half of the twentieth century has brought increased demands for governments of modern societies to expand their regulatory powers beyond economic to social regulation. As new technological hazards multiplied, traditional concerns with rates, routes, and pricing of industrial products and services were supplemented by pressures to safeguard the quality and safety of life. Issues such as public health, worker safety, medical devices, consumer protection, food production, and the environment either arose or gained in importance on the policy agendas of most industrial nations. With this shift, scientific knowledge became an ever more essential prerequisite for credible policy making, and governments vastly expanded their capacities for producing and assessing relevant technical information. The policy system's greatly enlarged dependence on science can be charted through the emergence in recent decades of new areas of research (e.g., environmental health, climate change), new analytic techniques (e.g., cancer risk assessment, biosafety assessment), and new programs of data collection (e.g., indicators for desertification or biodiversity, postmarket surveillance of adverse drug reactions).

As governments came to rely more on science as a basis for regulation, policy analysts initially assumed that cross-national cultural differences would diminish in importance as a factor shaping public action. The universality of science has been an article of faith for modern societies since the Enlightenment. A common base of scientific understanding, it was widely thought, would override the vagaries of national politics and culture in specific issue areas, whether nuclear power, pollution control, pharmaceutical regulation, the management of chemical pesticides, or the environmental release of genetically modified organisms (GMOs).

One of the more interesting findings of comparative policy research in recent years has been the failure of these expectations. Although policy *agendas*, broadly speaking, have converged on a host of issues worldwide, specific national *policies* for managing health, safety, and environmental risks continue to diverge, even when they are ostensibly based on the same bodies of scientific information. Intriguingly, evidence deemed persuasive in one national policy context does not necessarily carry the same weight in others. Even when policy outcomes converge, as for example in the informal moratorium on nuclear power across most of Europe and the United States, the underlying technical justifications are not invariably the same.

The literature on comparative policy provides some notable examples of cross-national divergences in the regulation of technological hazards. Thus, a four-country comparison of U.S. and European chemical regulation in the mid-1980s showed that European nations neither gave the same priority to carcinogens as did the United States nor developed comparable programs of testing and risk assessment (Brickman et al., 1985). Parallel differences have been observed even between the arguably more closely coupled policy systems of Canada and the United States (Harrison and Hoberg, 1994). National strategies for regulating air pollution have similarly diverged in priority setting, timing and severity of controls, and the choice of regulatory instruments. European countries, for example, were markedly slower to regulate airborne lead and chlorofluorocarbons than the United States. More recently, Europe has overtaken the United States in cutting sulfur emissions regarded as a precursor of acid precipitation. Biotechnological products created through genetic modification have encountered substantially different entry barriers on the two sides of the Atlantic, with significant cross-national disparities observable in the environmental release of GMOs (Jasanoff, 1995), the public acceptance of genetically modified foods, and patent protection for genetically engineered animals.

Numerous explanations have been offered for these persistent policy divergences, which reflect in turn underlying differences in societal perceptions and tolerance of risk. The simplest causal factor advanced by social scientists is economic interest—most plausibly invoked when the burdens and benefits of regulation fall disparately in different national contexts. For example, the relatively muted character of antinuclear protest in France (Nelkin and Pollak, 1981), as well as that nation's exceptionally low-key response to the Chernobyl disaster, have been attributed to the heavy French reliance on nuclear power as an energy source. Similarly, generators of acid precipitation such as the United States and the United Kingdom have been notably less aggressive in seeking control policies for sulfur oxides than the recipients of pollution, such as Canada and Norway.

Historical explanations seem to carry weight in other cases: Germany's unusual hostility to biotechnology in the 1980s no doubt reflected a distaste for state-sponsored science and a fear of uncontrolled genetic experimentation, both inherited from the Nazi era (Proctor, 1988; Gottweis, 1998). Other factors that have been held responsible for deviancies from allegedly rational policy choices include deficiencies in the public understanding of science (see, for example, Breyer, 1993), mass hysteria, the rise of politically influential social movements, the economic inefficiency of litigation (especially mass torts), and lack of political will or leadership. Cross-national variations in any of these factors could, in principle, lead to substantial divergences in policy outcome across countries.

The chief difficulty that these explanatory strategies encounter is their ad hoc and unsystematic character. Separate reasons are sought for each case of deviance from an idealized and supposedly rational baseline. Economics is invoked in one case, history in another, adversary politics in still a third. No general patterns emerge. Economic arguments, for all their appeal, only carry the day in a limited number of cases; often, national policies seem to favor outcomes that burden industrial production, as in the case of Europe's famous *precautionary principle* (see below) for the environment.

Moreover, sustained research in the fields of comparative policy and politics points to the durability of certain modes or styles of political action within nations, regardless of the issue in question. There appears to be a systemic quality in national responses to many different perceived threats and crises. The term *political culture* has been used as a catchall to explain such patterned divergences. It encompasses those features of politics that seem, in the aggregate, to give governmental actions a distinctively national flavor, even in countries sharing generally similar social, political, and economic philosophies. Political culture is difficult to measure in quantitative terms, although various attempts have been made to quantify some aspects of it. Thus, efforts were made in the 1960s to measure the engagement of citizens with their political system in several democratic societies (see, for example, Almond and Verba, 1963, 1989; Putnam, 1979, 1993), and many surveys have been made of public attitudes to particular technological developments (e.g., biotechnology, as measured by the Eurobarometer). For many, political culture is not a highly useful concept because it is simply the residue that remains after other efforts at rational explanation have failed. Yet it has become critically important to understand political culture better—especially with regard to its influence on public policy—as states, multinational corporations, and an increasingly well-informed civil society all confront the challenges of living together on the same bounded planet.

In this paper I attempt to synthesize our current knowledge of political culture as derived, empirically, from diverse studies of national regulatory systems and, theoretically, from recent developments in social theory and science and technology studies. In the following section of the paper I briefly outline the principal dimensions of variance among national approaches to regulating technological risks. In the subsequent section I outline the major ways in which comparative social scientists have tried to systematize the notion of political culture. In the concluding section I draw on this analysis to offer some reflections concerning the possible harmonization of policies for biotechnology across national boundaries.

DIMENSIONS OF CROSS-NATIONAL VARIANCE

Before embarking on a discussion of cross-national policy divergences, it is useful to remind ourselves of the large commonalities that provide humankind in any era with a shared set of experiences and understandings. In late modernity, as our historical moment is sometimes called, governments of advanced industrial societies have been required to deal with many common policy problems at roughly similar points in time. Examples with a significant scientific or technical dimension include, most recently, the global environmental crisis, the instability of global capital, economic restructuring after the Cold War, arms control, new epidemics, and the uneven social vulnerability to human as well as natural disasters. As one looks across the policy spectrum within liberal democracies, one finds numerous striking parallelisms in both governmental action and societal demand. These include similarities in legislative priorities, investments in science and technology, development of policy-relevant expertise, new forms of social mobilization, and increased interaction between state and nonstate actors. Significant shifts in policy ideology, such as economic liberalization or deregulation, are seldom any longer confined to the boundaries of single nations. Social movements, too, seem relatively unconstrained by national politics as they work to raise the visibility of particular policy issues. At the same time, national autonomy in many areas of policy making has been curtailed through increasingly thick networks of international regimes and institutions (for an overview of international developments in the environmental arena, see Haas et al., 1993).

Policy divergences are nested within this broad framework of common human understanding and social development. Their presence and persistence are therefore all the more remarkable. They deny any absolute claims for historical or technological determinism—that is, for universal regularities in human behavior occasioned by the characteristics of a particular period in time or by the material inventions of human ingenuity. Divergent responses to risk, in particular, point to the ability of social norms and formations—in short, of culture—to influence deeply the ways in which people come to grips with the uncertainties and dangers of the natural world. How do these different coping strategies most commonly manifest themselves?

Framing

Risk is often defined as the probability of a harmful consequence. How often will a flood or earthquake or volcanic eruption occur in a given region within a given number of years? What is the likelihood of an exceptionally severe El Niño or that warming of the earth's atmosphere will melt the West Antarctic ice shelf? What are the chances that a prolonged dry spell will give rise to consuming forest fires? Although risks of these types have traditionally been seen as natural, many risks of greatest concern to technological societies involve natural and social factors operating in tandem. How likely is it, for instance, that overfishing will destroy the capacity of fisheries to replenish themselves or that cutting down trees for fuel will lead to deforestation of uplands and consequent downstream flooding? With increasing knowledge of human–nature interactions, we have come to perceive numerous phenomena once seen as wholly natural as also having a human-made component. Anthropogenic climate change is perhaps the most noteworthy example of such a shift in awareness.

It is by now widely acknowledged in the policy analytic literature that our capacity to identify distinct problems from a universe of potentially interconnected causes and effects involves a kind of selective vision referred to as *framing* (see, for example, Cobb and Elder, 1972; Dryzek, 1990; Schon and Rein, 1994). Frames have been defined as "principles of selection, emphasis, and presentation composed of little tacit theories about what exists, what happens, and what matters" (Gitlin, 1980). By sorting experience into such well-demarcated patterns of significant causes and effects, human agents impose meaning on what might otherwise be no more than a jumble of disconnected events. Framing orders experience, eases confusion, and creates the possibility of control. Problems that have been framed are capable, in principle, of being managed or solved. At the same time, framing, in its nature, is also an instrument of exclusion. To bring some parts of experience within a frame—to render them comprehensible and interpretable—other parts must be left out as irrelevant, incomprehensible, or uncontrollable. This dual aspect of framing, as a device for ordering as well as exclusion, helps to capture some of the observed cross-cultural variation in the identification and management of risk.

Differences in framing are most starkly apparent when the same social problem is attributed to different causes by competing actors in a policy-making environment. Is climate change the result of worldwide emissions of greenhouse gases, as claimed by Western scientists, or is it the result of centuries of unsustainable and inequitable resource exploitation by industrial countries, as claimed by some developing country activists? Is the world as we know it teetering on the brink of environmental disaster because of overpopulation in poor countries or overconsumption in rich ones? Should the AIDS epidemic be seen as a consequence of deviant sexual behavior or is it simply a highly resistant viral disease that foreshadows the threat of new global epidemics? Is persistent poverty attributable to welfare policies that sap individual initiative or to the absence of effective job creation strategies in inner cities? Opinion on such questions may differ radically among actors within a single country as well as between countries. For purposes of this paper, differences of the latter type are of greater interest because they point to the possible influence of culture rather than of more temporary economic or social interests.

National responses to the risks of biotechnology in recent decades provide one striking example of divergent framings of technological risk (see Jasanoff, 1995 for further details). In the United States, initial concerns with the safety of genetic engineering as a new scientific process gave place to a government-wide consensus that regulation should focus largely on the *products* of biotechnology. The process of genetic manipulation was deemed not to pose any special hazards in and of itself. In Britain, by contrast, policy leaders have continued to worry about genetic engineering as a novel *process* that is not well enough understood to be granted a clean bill of health. In Germany, the risks of biotechnology were seen from the start as both social and natural in character, because the technique appeared to give the state unregulated power to reshape the meanings of nature and human identity. The uncertain risks of genetic research seemed to undercut the German constitutional system's guarantee of adequate state protection against industrial hazards. Accordingly, Germans felt the need for *programmatic* legislation in the form of a law specifically addressing genetic engineering to control this technological enterprise in all its aspects.

Styles of Regulation

Although governments of industrial societies frequently converge in deciding which risks require positive state action, resulting regulatory programs are often founded on very different patterns of interaction between the state and other major actors. These systematic differences are sometimes referred to as *styles* of regulation (Vogel, 1986; Brickman et al., 1985). The components of a nation's regulatory style may include, in brief, the means by which the state solicits input from interested parties, the opportunities afforded for public participation, the relative transparency of regulatory processes, and the strategies employed for resolving or containing conflict.

Comparative research over the past two decades has highlighted the relatively sharp stylistic differences between the United States and other industrial countries. On the whole, U.S. regulatory processes are more formal in soliciting and processing information, more inclusive in securing participation, more comprehensively documented, and more adversarial in handling disputes than those of most other nations. Thus, U.S. administrative law permits private parties to sue regulators for both substantive and procedural deficiencies in their decision making; agencies may therefore be sued for failure to build an adequate technical record or to take account of relevant scientific information. In other countries, litigation against regulators is at best infrequent and even then is limited to instances in which a right has been clearly violated. Lawsuits founded on alleged inadequacies in the government's technical analysis, such as the recent appeals court decision striking down proposed federal standards on airborne particulates (DC Cir., 1999), are virtually unheard of outside the United States. Disputes elsewhere are resolved more often behind closed doors than in the open forum of a courtroom. Correspondingly, the basis for policy decisions is far less readily available to the public at large than in the United States, where the Freedom of Information Act and other laws create a strong presumption in favor of disclosure.[1]

These generic differences in communications between state and society also intersect with the manner in which social actors express dissent from, or resistance to, official policy. Discontent with state action manifests itself most readily in the United States, where the courts provide a ready avenue for challenging policy decisions. Highly polarized conflicts are more likely to give rise to direct political action in Europe and Japan, as when Greenpeace occupied the Brent Spar oil platform off the British coast or when various environmental groups have blockaded construction sites, torn up fields planted with GMOs, and the like. In the United States, comparably sharp disagreements would far more probably end up in court. Direct actions, such as terrorist attacks by animal rights activists or by the Unabomber, are considered here the exceptions, not the norm. Public referenda, widely used in a number of smaller European states, are atypical in U.S. politics, although the state of California has been a notable outlier in this respect.

Acceptable Evidence

Differences in the framing and style of regulation go hand in hand with substantial differences in the kinds of evidence that governments and the public consider suitable as a basis for public decisions. Standards of proof and persuasion also differ across countries, along with preferences for particular methods of technical analysis.

Contrasts between the United States and major European countries again provide some of the most striking examples. Comparative researchers have noted the consistent U.S. preference for formal and quantitative analytic methods, whether in measuring risk, economic costs and benefits, or even the relatively intangible impacts of regulatory policy on social justice. U.S. environmental policies, for example, gave highest priority throughout the 1970s and 1980s to the risk of chemically induced cancer. During this time, significant energy went into the development of sophisticated analytic techniques designed to produce reliable quantitative estimates of risk and, eventually, the uncertainty surrounding such estimates (NRC, 1994). European countries facing presumably comparable problems avoided the use of formal quantitative techniques in favor of more qualitative appraisals based on the weight of the evidence (Jasanoff, 1986).

These differences in forms and standards of acceptable evidence appear to correlate well with the two terms of greatest legal significance—*risk* and *precaution*—that have helped to define preventive environmental policy making

---

[1]A recent manifestation of the bias toward openness was the inclusion of a directive in the 1999 omnibus spending bill requiring the Office of Management and Budget to amend its rules for extramural research grants so that "all data" collected using federal research funds would be accessible under the Freedom of Information Act.

during the past two decades. *Risk*, as already noted, is the term heavily favored in U.S. legislation and public policy, whereas European nations have tended to attach greater consequence to the *precautionary principle*. These terms not only reflect subtly different notions about the purpose and scope of environmental protection, but they also entail different approaches to the public justification of environmental policy.

The concept of risk appears at first glance to render environmental problems more tractable precisely because it is probabilistic and measurable. The term was borrowed into the environmental domain from the financial sector, where it refers to a quantifiable probability of one or another adverse outcome. Risk is actuarial in spirit. One can (indeed, one often *must*) insure oneself against various kinds of risks for which actuarial data are available, such as fires, floods, earthquakes, catastrophic illnesses, or automobile accidents. When used in environmental decision making, risk retains the connotation of something that can be clearly defined and quantified, hence managed. It is a relative concept that risks can always be offset against benefits, and risk-based laws often explicitly prescribe that the benefits of policy action (which are, in their turn, quantified) should outweigh the risks. Importantly as well, risks can be compared against one another (Graham and Wiener, 1995) so that policymakers can be meaningfully instructed to focus on large risks over small ones and to ignore altogether risks that are too tiny to matter.

Critics of risk-based policy have noted that the language of risk implicitly conceptualizes most human-environment interactions as harmless or even positively beneficial (Winner, 1986). Risk is thought to be the exception, not the rule, in human engagements with nature. It is something that one can guard against without upsetting underlying philosophies of development, consumption, or resource use. By comparison, the precautionary principle seems to display greater sensitivity to human ignorance and uncertainty. Historically, the term is a translation of the German *Vorsorgeprinzip*, one of five fundamental principles recognized in German law as constituting the basis for environmental policy. Migrating into the English language and into European policy, the term has inevitably lost precision, but some of its features are quite generally accepted. The principle states in brief that damage to the environment should be avoided in advance, implementing a duty of care on the part of policymakers. As with risk, the principle emphasizes prevention rather than cure. But precaution, as used in a wide variety of European policy statements, seems to urge something more than mere prevention. It demands heightened caution in the face of uncertainty, to the point of favoring inaction when the consequences of action are too unclear. Unlike risk, which both invites and lends itself to calculation, precaution implies a greater need for (uncalculated) judgment and, where necessary, restraint.

Precaution, to be sure, is never an absolute mandate in any regulatory system. Just as risks are balanced against benefits, so the precautionary principle is offset in practice by other moderating principles, such as the requirement that actions be proportional to the anticipated harm. Nonetheless, the very indeterminacy of the idea of precaution may have kept it from being translated into formal assessment methodologies, such as quantitative risk assessment or risk-benefit analysis. Put differently, it may be easier to work with a concept such as precaution in a cultural environment that does not insist on mathematical demonstrations of the rationality of policy decisions. The preference for relatively formal and quantitative or relatively informal and judgmental techniques of decision making thus resonates with other important values in environmental policy making.

Forms of Expertise

Whether formal or informal, risk analytic frameworks incorporate tacit assumptions about how the world works; the use of analytic techniques, moreover, entails choices about who participates, and how, in processes of environmental decision making. Both the forms of relevant expertise and the rules of participation may differ substantially in national systems for dealing with the same regulatory problems.

One axis of divergence that has proved to be especially significant is whether experts are selected primarily on the basis of their technical qualifications (*what* they know) or as much on the basis of their institutional affiliations and experience (*who* they know, and in what context). On the whole, the U.S. policy process stresses experts' technical competence more than their institutional or political background. American regulatory statutes not infrequently specify, for instance, which types of technical expertise must be represented on advisory panels for federal agencies. In many U.S. policy frameworks, expert advisers are actively required to display their political independence and neutrality as a prerequisite for government service. Although most such bodies also have to meet requirements of breadth and inclusiveness, overt application of political criteria is deemed inappropriate in most cases (for examples and further discussion, see Jasanoff, 1990). Objective scientific expertise is generally valued more highly than other grounds for decision making, and attacks on the scientific competence of regulatory agencies is a standard device for undermining their political legitimacy.

By contrast, expert advisory bodies in other industrial nations are often more explicitly representative of particular interest groups and professional organizations. Tripartite arrangements, including government, industry, and labor, are especially commonplace; in newer regulatory frameworks, participation has sometimes been broadened to include representatives of social movements, such as environmentalists and consumers. Outside the United States, an expert body thus is thought to reflect in microcosm the segment of society which will be affected by its policy advice. Expert judgment is expected to be binding because the group as a whole is capable of speaking for the wider community it represents. Technical expertise, experience or tacit knowledge, and social identities are regarded as equivalently important qualifications for offering advice under these presuppositions.

Another important dimension of difference concerns the role of nonexpert opinions in decision-making. Again, the U.S. policy system is most open to the inclusion of such viewpoints in the decision making record. Formally, inclusiveness is assured through a process that offers interested parties, at a minimum, the chance to comment on the government's rationale for proposed decisions. In many areas of social regulation, lay participation is secured through more formal means, such as administrative hearings that give nonexperts a chance both to present their own positions and also to question those of technical experts representing government and industry. Ordinarily, entry into the U.S. policy process occurs at the initiative of parties who see themselves as stakeholders. By contrast, in most European countries, the right to be recognized as a stakeholder is neither automatic nor achieved through self-selection, but must be officially acknowledged through legislation or administrative practice. Entry accordingly tends to be limited to groups or actors who have established longstanding or politically salient working relations with governmental agencies.

Nature of Regulatory Standards

Standards play a crucially important role in any policy system that seeks to protect the public against technological risks. Standards come in many forms. They may be applied to industrial processes, pollutants, facilities, products, equipment or vehicles, or natural media such as air and water. Standards may be used to regulate the quality of an environmental medium; control harmful discharges, emissions, and residues; establish limits for human exposure to toxic substances; specify safe usage conditions for regulated products; or influence environmentally detrimental behaviors. In effectuating these goals, standards may directly address the design of a product or process (e.g., air bags in automobiles) or specify the performance level desired of a particular technology, leaving the means of compliance more flexible (e.g., emissions standards for power plants). They may be required by law (regulatory standards), recommended by guidelines, or voluntarily adopted by industries or private standard-setting organizations (consensus standards). They may be enforced through rigorous governmental monitoring and legal sanctions or through economic incentives or through relatively lax systems of self-regulation.

From this wide range of possible variation, national policy systems often seek out some characteristic approaches to standard setting. U.S. regulation, for example, has shown a preference since the early 1970s for nationally uniform standards that are enforced through the legal process. Penalties can be harsh, sometimes in the form of criminal sanctions for corporate executives. At the other extreme, British environmental standards were at one time locally and flexibly negotiated to suit the economic and technical capabilities of particular industrial concerns. More recently, this national preference has yielded somewhat in the face of European demands for greater uniformity and accountability across member states. With respect to enforcement, the European approach overall is less adversarial and legalistic than the American approach. Compliance tends to be achieved through bargaining and behind-the-scenes negotiation between business and government (other social actors generally play little role in enforcement) rather than through the Draconian processes of the law.

VARIETIES OF CULTURAL EXPLANATION

That differences such as those described above persist across similarly situated societies has presented a puzzle to economics and political science. Cross-national variations in risk perception and risk policy appear to contradict widely held assumptions of technical as well as economic rationality, both of which would predict greater convergence when states act upon similar information and need to balance similar trade-offs between the benefits and burdens of regulation. To explain patterned divergences in societal responses to risk, one has to supplement theories of rational choice with approaches that focus more centrally on the public interpretation of experience—in short, to supplement studies of reason and utility with studies of culture and meaning. In particular, one has to examine the role of institutions in stabilizing particular ways of dealing with uncertainty, conflict, expertise, and participation.

Going beyond currently dominant theories that cast states as rational actors, comparative studies of risk have given rise to three main theoretical frameworks for understanding cultural variations. The first is *structural*. This approach places primary emphasis on the role of political organization. It is presumed that the ways in which power is formally divided in society profoundly influence public perceptions of security and insecurity and also channel governmental action in specified directions. The second framework is *functional*. This approach regards all societies as encountering recurrent problems in the form of threats to their welfare or existence. Functionalist explanations therefore tend to see cross-cultural policy variations as by-products of differences in the perception, or framing, of social problems among different societies. The third framework is *interpretive*. This approach places primary emphasis on the need of societies to make sense and meaning of their collective experience, taking into account changes in knowledge and human capacity produced through science and technology. Interpretive social theorists—including specialists in science and technology studies—are particularly interested in the instruments of meaning creation in society, including most importantly various forms of language or discourse. Each framework illuminates some of the causes of cultural variation in risk perception and risk policy, as briefly described below.

The Role of Political Structure

The ways in which governmental power is institutionalized influence a society's handling of risk in more or less obvious directions. At the simplest level, agencies that are responsible for both the promotion and the regulation of technology tend to be more accepting of risk than those whose mandate is limited to regulation. This is why, in 1972, U.S. environmentalists successfully pressed to have pesticide regulation removed from the Department of Agriculture, where agribusiness interests were considered dominant, to the newly formed and politically less committed Environmental Protection Agency. Similarly, the regulation of nuclear power was taken away from the old Atomic

Energy Commission and delegated to the more independent Nuclear Regulatory Commission. Failure to separate promotional and regulatory functions in this way arguably leads to laxer regulatory practices. For instance, Britain's Ministry of Agriculture, Fisheries and Food is widely thought to have underestimated the transmissibility of bovine spongiform encephalopathy (BSE or mad cow disease) because its primary goals were to help the beef industry and prevent public panic.

More generally, the institutional organization of power affects the ways in which nongovernmental actors fight for particular policy objectives. In parliamentary democracies, for example, electoral politics provides the primary avenue through which citizens can expect to influence government. The rise of Green parties in Europe illustrates this dynamic. Environmentalists have needed to muster seats in parliament in order to press their agendas, and their success rates have differed from one country to another. In the United States, by contrast, local groups such as NIMBY ("not in my backyard") organizations have largely taken the place of party politics. The readiness of American citizens to form single-issue associations is historically documented. This strategy is facilitated by a "political opportunity structure" in which power over many issues is decentralized and local initiative can express itself through a variety of mechanisms, such as lawsuits, local referenda, and elections. In countries with more hierarchical and closed decision-making processes, activist groups may be slower to form and there may be an appearance of greater trust in government. However, underlying such superficial political acquiescence there may be significant public alienation and distrust which can erupt into mass protest if the opportunity arises (for sociological accounts of such public attitudes in Europe, see Beck, 1992; Irwin and Wynne, 1996).

Structural divisions of power have been plausibly correlated with another aspect of national regulatory styles, namely, the degree to which decisionmakers rely on formal, objective, or quantitative justifications for their actions. In the relatively transparent and competitive U.S. policy environment, decisionmakers are apt to be vulnerable to charges of subjectivity and arbitrariness. Indeed, the federal Administrative Procedure Act authorizes courts to review agency decisions to ensure that they are not arbitrary or capricious. Given these pressures, it is not surprising that United States policymakers have opted over time for more explicit and formal analytic techniques than their counterparts in other advanced industrial states. Examples include quantitative risk assessment of chemical carcinogens, cost-benefit analysis of proposed projects, detailed economic analysis of regulatory impacts, and environmental equity analysis— all of which are more extensively used, and also debated, in the United States than in other liberal democracies.

For all their apparent power, structural explanations have some notable deficiencies. Because they take structures for granted, they are unable in principle to account for modification and change in institutional configurations of power. Some phenomena that have proved important in international risk debates but that elude structural analysis include the rise and transnational spread of social movements and epistemic communities (groups of actors sharing similar beliefs and values about a given issue area), the shifts from one problem framing to another in national and international programs of risk management, and the differences in value commitments with respect to technological risk among similarly situated states and social actors. Other types of cultural explanation have proved more helpful for these purposes.

The Functionalist Approach

Functionalist approaches, as noted above, conceive of societies as having a range of large problems that continually need to be addressed and solved for the society's general well-being. Risk could be seen as one such problem. Unmanaged risk creates situations of extreme uncertainty for citizens and undermines confidence in ruling institutions. Social theorists have argued that the rise of modern regulatory states was in part an answer to the risks of widespread economic and social dislocations surrounding the industrial revolution. In particular, institutions such as the insane asylum and the workhouse and social analytic techniques such as statistics and demography are thought to be instruments developed by states in order to enable and maintain policies for social order (see, for instance, Foucault, 1979; Porter, 1986; Nowotny, 1990).

One of the best known attempts to understand cultural variations in the management of risk arises from a blending of anthropology and political science in work initiated by Douglas and Wildavsky (1982). Cultural theorists have noted that beliefs about nature and society are encountered in some commonly recurring clusters that appear to correlate with forms of social organization. Three dominant belief systems about environmental problems have been described most often in the literature: catastrophist or preventivist (nature is fragile); cornucopian or adaptivist (nature is robust); sustainable developmentalist (nature is robust within limits) (Cotgrove, 1982; Jamison et al., 1990; Rayner, 1991). The image of nature as cornucopian has been further differentiated into the idea that natural bounty is lottery-controlled cornucopian (nature is capricious) or else that it is freely available (nature is resilient) (Thompson et al., 1990). Cultural theory posits that these persistent forms of belief are not accidental but are connected to underlying features of social order.

To explain why human views of nature, and associated views about human nature, fall into certain broad patterns, cultural theory suggests that such beliefs grow out of a need to preserve important ordering elements in social relations. Douglas (1970), in particular, sees two cultural variables as fundamental: hierarchy within a community ("grid") and the firmness of its demarcation from other communities ("group"). For example, bureaucratic organizations (high grid and high group, in Douglas' terms) are most inclined to believe that nature, though not infinitely malleable, can be managed by means of appropriate, technically grounded, and formally legitimated rules. Such beliefs promote this culture's interest in protecting its boundaries against outsiders, as well as in preserving its clear internal hierarchy. In contrast, market or entrepreneurial cultures (low grid and low group) seem more likely to subscribe to a cornucopian view of nature—that is, the capacity of nature to rebound from assaults without active human intervention. This belief is consistent with the culture's willingness to rearrange its membership and operating rules so as to make best use of changes in its environmental resources.

By reducing the complexity of human–nature interactions to a few fixed types, the categories of cultural theory run up against some significant theoretical difficulties. It is unclear, to begin with, whether so parsimonious a notion of culture can be applied in meaningful ways to complex organizations (firms, social movements), let alone to nation states. Moreover, institutions and their members appear in this framework to be inflexibly bound together in hard and fast belief systems. This rigid packaging contradicts the ambivalence and heterogeneity of response reported in the literature on risk perception and public understanding of science and technology.

Cultural theory also resembles structural approaches in its relative insensitivity to historical processes. A functionalist notion of culture tends to take the needs of particular cultural types for granted. A bureaucracy, for instance, is *always* looking to maintain its hierarchical integrity, just as entrepreneurs are *always* seeking to maximize their profits through new modes of resource exploitation. Such assumptions are not well suited to account for large-scale social and ideological movements, such as the shift in the Western world from a pollution-centered to a sustainable developmentalist philosophy of environmental management in the 1980s. Shifts within organizations are also puzzling from the standpoint of cultural theory. For example, why was there a "greening" of industry in the late twentieth century, and why did German environmentalists eventually drop their "just say no" stance toward biotechnology? Changes in scientific understanding could provide part of the answer in such cases, but science, technology, and expertise play a relatively passive or subordinate role in the cultural theory framework. Science is seen more as a resource to be controlled by the dominant cultural types than as a source of distinctive knowledge and persuasive power. Nonetheless, cultural theory valuably calls attention to the socially constructed character of beliefs about nature and to possible connections between longstanding social relations and the perception and management of risk.

Interpretive Approaches

Interpretive social theory focuses from the start on the place of ideas in social life. It asks how people make sense of what happens to them, how they distinguish between meaningful and meaningless events, and how they accommodate themselves to new information or experience. It regards culture as the lens through which people understand their condition. This approach is centrally concerned with the origins of and changes in belief systems, including the modern belief system called science, and with the factors that make certain beliefs either unquestionable (ideology) or else massively resistant to modification. Accordingly, interpretive work in the social sciences has focused on the resources with which societies construct their ideas, beliefs, and interpretations of experience. These include aspects of social behavior that have not been widely examined in quantitative social sciences, such as language and visual representation.

An important contribution of this theoretical approach has been to show how formal systems of language and practice incorporate particular, often culturally specific, ways of looking at the world—in other words, how they help to frame both problems and solutions (see, for example, Bjork, 1992; Litfin, 1994). Quantitative risk assessment (QRA) of chemicals provides an especially instructive example for our purposes. As noted above, this analytic technique has been more extensively used in the United States than in other industrial countries. Its use, in turn, implies a number of prior assumptions about the nature of risk and uncertainty.

QRA builds not only on seemingly objective measurements of toxicity and exposure but also, less visibly, on underlying models of causality, agency, and uncertainty. It frames the world, so that users of the technique are systematically alerted to certain features of risk but desensitized to others. Causation for purposes of QRA, for example, is generally taken to be simple, linear, and mechanistic. Asbestos causes cancer and dioxin causes birth defects in animals, but perhaps not in humans. The classical model of cancer risk assessment used by most U.S. federal regulatory agencies conceives of risk as the result of individual or population exposure to single harmful substances. Over the years, this causal picture has grown in complexity. An older single-hit model of carcinogenesis has been replaced by one that views cancer as a multistage process. It is recognized as well that risk is distributed over populations of varying composition and susceptibility, exposed for variable lengths of time and by multiple pathways. Quantitative models have been redesigned to reflect these discoveries.

But a closer look reveals that some of the most up-to-date models of risk assessment still remain quite partial and selective in their treatment of causes. In focusing on particular substances, for example, QRA necessarily ignores others. Despite scientific arguments to the contrary, industrial chemicals are taken to be of greater public health concern than similar substances to which people are exposed by nature. QRA in this way treats causes as if they fall primarily on the artificial, or non-natural, side of human exposure to chemicals in the environment (Ames et al., 1987; Gold et al., 1992).

In other respects, QRA tends to simplify the world so as to dampen the overall estimate of risk. The impact of multiple exposure routes and possible synergistic effects, for example, is rarely captured. Behavioral patterns that may aggravate risk for particular subpopulations (a well-known example is smoking among asbestos workers) are similarly downplayed or disregarded. Socioeconomic factors that tend to concentrate risk from many sources for poor and minority populations were not normally considered in QRA until pressure to do so arose from the environmental justice movement.

QRA also incorporates tacit conceptions of agency. Implicit in this mode of analysis is the notion that risk originates in the inanimate world, even though it is known at some level that social behavior is part of the process that produces risk. By focusing on material agents as the primary sources of risk, QRA tends to diminish the role of human agency and responsibility. An indirect consequence is that governmentally sponsored research on risk has centered around issues of concern to mathematical modelers rather than to social scientists more broadly. Yet organizational sociologists have observed for years the complex ways in which the physical and human elements of technological systems interact to produce risky conditions and disasters (see, for example, Perrow, 1984). Similar insights have emanated as well from the sociology of technology, which calls attention to the continual interplay of functions between animate and inanimate actors (Bijker et al., 1987).

The third set of assumptions embedded in QRA has to do with the nature of uncertainty and our perceptions of it. This method takes for granted that it is possible to encapsulate in objective and understandable forms the zones of uncertainty that regulators should be aware of when attempting to control risk. Comparative and historical work on risk has shown, however, that even when societies use quantitative analysis to further public policy, they differ in how they classify and measure natural phenomena, which techniques they label as objective or reliable, how they characterize uncertainty, and what resources they apply to its reduction (Porter, 1995). Far from being a neutral statement about the unknown, uncertainty about risk thus appears as the product of culturally situated forms of activity. It is a collectively endorsed recognition that there are things about our condition that we do not know; but such an admission is only possible because there are agreed-upon mechanisms for finding out more.

QRA users, and quantitative modelers more generally, will tend to think about causation, agency, and uncertainty in different terms from those who rely on qualitative approaches to risk assessment. In European decision-making environments, for example, the interconnectedness of social and natural causes may be more readily understood, provided that policy advisory bodies include a sufficiently diverse range of expertise. Uncertainty is managed by building trust in particular institutions rather than by expressing it more precisely through formal analytic techniques. Thus, British policy has historically relied on a tested cadre of public servants whose integrity and judgment are considered beyond doubt (Jasanoff, 1997). German policy, by contrast, depends to a large extent on agreements forged in consensual, politically representative expert bodies whose decisions are trusted because they reflect the full spectrum of relevant societal beliefs. Uncertainty within these contexts is most likely to manifest itself as a loss of trust in the experts or expert bodies responsible for making policy.

QRA for its part loses credibility when it openly ignores spheres of human experience that bear crucially on people's perception of risk (for case studies of such loss of trust, see Krimsky and Plough, 1988). These include the strength of family and work relationships, the robustness of communities, the special status of children, and the trustworthiness of major institutions. Failure to take account of such historical and cultural factors in risk determinations can induce alienation, distrust, and heightened risk perception in those who are unable to participate meaningfully in the preserves of objective technical expertise. These observations account for recent high-level recommendations in U.S. policy circles to interweave processes of technical analysis and political deliberation more closely in risk decision making (NRC, 1996; Presidential/Congressional Commission, 1997).

CONCLUSIONS

Differences in institutionalized divisions of power, culturally grounded perceptions of need, and formalized systems of analysis profoundly shape the ways in which technological risks are framed for purposes of policy making. Contradicting the expectations of rational choice and policy convergence, these factors produce divergences in the conceptualization and management of risk even among societies that are closely similar in their economic, social, and political aspirations. As risk debates are globalized, engaging vastly more disparate societies, one can only expect such divergences to harden and grow more numerous. Cultural differences are particularly likely to arise when a risk domain touches upon issues that are basic to a society's conceptions of itself, such as constitutional relations between science and the state or religious and philosophical ideas about what is "natural." What then are the implications for the future of a promising new technology, especially one such as biotechnology that impinges upon such a wide range of fundamental conceptual questions?

One source of optimism is the proliferation in recent decades of policy-harmonizing institutions in the international arena. Their existence, and the increasing scope and diversity of their mandates, testify to the desire of modern societies to progress toward a shared future of increased safety, health, material comfort, and psychological well-being. Yet in trying to meet these multiple demands, international harmonizing bodies risk falling victim to the so-called contradictions of postmodernity. Different cultural constructions of the "same" policy problem may make agreement difficult in spite of apparent similarities in national goals and aspirations. Even where consensus is reached, ambiguities may subsequently resurface in the process of implementation. An initial convergence among experts may not always be sufficient to reassure skeptical publics and ensure robust political acceptance.

The 1996 BSE scare in Europe provided a dramatic but typologically by no means isolated example. The European Union's efforts to construct a unified, science-based standard to calm citizens confronting (ostensibly) the "same" risk of disease from the "same" agent were undercut by the discrepant perceptions of farmers, parents, food producers, government scientists, independent scientists, public health officials, agriculture ministers, politicians facing reelection, anti-European Britons, and the Brussels bureaucracy. Quantitative analysis proved inadequate for bridging these far-flung interests, as ministers wrestled week after week to agree on a single magic number—the number of cows that would have to be culled to render the beef supply adequately safe for all uses. Cartoons, black humor, and bizarre role reversals took the place of orderly policy making. Butchers in the markets of Europe appropriated the expert's reassuring role, with official-looking signs to back up their guarantees of "no British beef sold here." Ministers, having vainly turned to science for credibility, were forced to regain trust through personalized expressions of consumer confidence, such as, "Beef will still be served. Myself and my family will continue to eat beef" (U.K. Minister John Gummer, as quoted in the *Independent*, March 22, 1996, pg. 5; see also Jasanoff, 1997).[2]

National policy institutions—shored up by history, tradition, established policy discourses, and well-understood standards of fairness and rationality— are able to persuade most of their publics most of the time that they can deliver fair and objective solutions to complex problems. International harmonizing bodies have few if any of these legitimating props at their disposal. As risks such as BSE assume global proportions, harmonizing institutions are likely to find it difficult to pass off as impartial expert judgment the political act of mediating among competing cultural framings of risk. Yet international regulatory institutions remain for the most part less transparent and less accessible to public input than their counterparts within many national governments.

Letting politics back into international policy processes may therefore be more productive in many cases than leaning exclusively on the supports of allegedly rational policy analysis. Mutual education seems the most promising route to eventual cross-national harmonization. If culture permeates the ways in which people cope with risk, then learning to understand each other's framing processes becomes a necessary prelude to collective action in the international arena. Exploring how culture matters in the politics of risk constitutes a modest first step in this direction.

REFERENCES

Almond, G.A., and S. Verba. 1963. The Civic Culture: Political Attitudes and Democracy in Five Nations. Princeton, N.J.: Princeton University Press.

Almond, G.A., and S. Verba, eds. 1989. The Civic Culture Revisited. Newbury Park, Calif.: Sage Publications.

Ames, B.N., R. Magaw, and L.S. Gold. 1987. Ranking Possible Carcinogenic Hazards. Science 236:271–80.

Beck, U. 1992. Risk Society: Towards a New Modernity. London: Sage Publications.

Bijker, W.E., T.P. Hughes, and T. Pinch, eds. 1987. The Social Construction of Technological Systems. Cambridge, Mass.: Massachusetts Institute of Technology Press.

Bjork, R. 1992. The Strategic Defense Initiative: Symbolic Containment of the Nuclear Threat. Albany: State University of New York Press.

Breyer, S. 1993. Breaking the Vicious Circle: Toward Effective Risk Regulation. Cambridge, Mass.: Harvard University Press.

Brickman, R., S. Jasanoff, and T. Ilgen. 1985. Controlling Chemicals: The Politics of Regulation in Europe and the U.S. Ithaca, N.Y.: Cornell University Press.

Cobb, R.W., and C.D. Elder. 1972. Participation in American Politics: The Dynamics of Agenda-Building. Baltimore, Md.: Johns Hopkins University Press.

---

[2]A new public crisis surrounding genetically modified foods that broke loose in Britain in February and March 1999 echoed many of the same themes. Prime Minister Tony Blair appeared to have learned little from the BSE episode as he tried to reassure Britons by saying that he personally would be happy to consume genetically modified foods.

Cotgrove, S. 1982. Catastrophe or Cornucopia: The Environment, Politics and the Future. Chichester, U.K.: Wiley.

DC Cir. 1999. American Trucking Associations, Inc., et al. v. United States Environmental Protection Agency, No. 97-1440.

Douglas, M. 1970. Natural Symbols: Explorations in Cosmology. London: Barrie and Rockliff.

Douglas, M., and A. Wildavsky. 1982. Risk and Culture. Berkeley: University of California Press.

Dryzek, J.S. 1990. Discursive Democracy—Politics, Policy, and Political Science, Cambridge, U.K.: Cambridge University Press.

Foucault, M. 1979. Discipline and Punish. New York: Vintage.

Gitlin, T. 1980. The Whole World Is Watching: Mass Media in the Making and Unmaking of the New Left. Berkeley: University of California Press, p. 6.

Gold, L.S., N.B. Manley, and B.N. Ames. 1992. Extrapolation of Carcinogenicity Between Species: Qualitative and Quantitative Factors. Risk Analysis 12.

Gottweis, H. 1998. Governing Molecules: The Discursive Politics of Genetic Engineering in Europe and the United States. Cambridge, Mass.: Massachusetts Institute of Technology Press.

Graham, J.D., and J.B. Wiener, eds. 1995. Risk versus Risk. Cambridge, Mass.: Harvard University Press.

Haas, P.M., R.O. Keohane, and M.A. Levy. 1993. Institutions for the Earth. Cambridge, Mass.: Massachusetts Institute of Technology Press.

Harrison, K., and G. Hoberg. 1994. Risk, Science, and Politics: Regulating Toxic Substances in Canada and the United States. Montreal: McGill-Queen's University Press.

Irwin, A., and B. Wynne, eds. 1996. Misunderstanding Science? Cambridge, U.K.: Cambridge University Press.

Jamison, A., R. Eyerman, and J. Cramer. 1990. The Making of the New Environmental Consciousness; A Comparative Study of the Environmental Movements in Sweden, Denmark, and the Netherlands. Edinburgh: Edinburgh University Press.

Jasanoff, S. 1986. Risk Management and Political Culture. New York: Russell Sage Foundation.

Jasanoff, S. 1990. The Fifth Branch: Science Advisers as Policymakers. Cambridge, Mass.: Harvard University Press.

Jasanoff, S. 1995. Product, process, or programme: three cultures and the regulation of biotechnology. In M. Bauer, ed., Resistance to New Technology. Cambridge, U.K.: Cambridge University Press.

Jasanoff, S. 1997. Civilization and madness: the great BSE scare of 1996. Public Understanding of Science 6:221–232.

Krimsky, S., and A. Plough. 1988. Environmental Hazards: Communicating Risks as a Social Process. Dover, Mass.: Auburn House.

Litfin, K.T. 1994. Ozone Discourses: Science and Politics in Global Environmental Cooperation. New York: Columbia University Press.

NRC (National Research Council). 1994. Science and Judgment. Washington, D.C.: National Academy Press.

NRC (National Research Council). 1996. Understanding Risk: Informing Decisions in a Democratic Society. Washington, D.C.: National Academy Press.

Nelkin, D., and M. Pollak. 1981. The Atom Besieged. Cambridge, Mass.: Massachusetts Institute of Technology Press.

Nowotny, H. 1990. Knowledge for certainty: poverty, welfare institutions and the institutionalization of social science. In P. Wagner, B. Wittrock, and R. Whitley, eds. Discourses on Society 15:23–41.

Perrow, C. 1984. Normal Accidents. New York: Basic Books.

Porter, T.M. 1986. The Rise of Statistical Thinking 1820–1990. Princeton, N.J.: Princeton University Press.

Porter, T.M. 1995. Trust In Numbers: The Pursuit of Objectivity in Science and Public Life. Princeton, N.J.: Princeton University Press.

Presidential/Congressional Commission on Risk Assessment and Risk Management. 1997. Framework for Environmental Health Risk Management. Washington, D.C.: Presidential/Congressional Commission.

Proctor, R. 1988. Racial Hygiene: Medicine under the Nazis. Cambridge, Mass.: Harvard University Press.

Putnam, R.D. 1979. Studying elite political culture: the case of ideology. American Political Science Review 65:651.

Putnam, R.D. 1993. Making Democracy Work: Civic Traditions in Modern Italy. Princeton, N.J.: Princeton University Press.

Rayner, S. 1991. A cultural perspective on the structure and implementation of global environmental agreements. Evaluation Review 15(1):75–102.

Schon, D.A., and M. Rein. 1994. Frame/Reflection. New York: Basic Books.

Thompson, M., R. Ellis, and A. Wildavsky. 1990. Cultural Theory. Boulder, Colo.: Westview Press.

Vogel, D. 1986. National Styles of Regulation. Ithaca, N.Y.: Cornell University Press.

Winner, L. 1986. On not hitting the tar-baby. Pp.138–154 in The Whale and the Reactor: A Search for Limits in an Age of High Technology. Chicago, Ill.: University of Chicago Press.

Part II

Political and Ecological Economy

.

5

Biological Impacts of Species Invasions: Implications for Policymakers

KAREN GOODELL
Department of Ecology and Evolution,
State University of New York at Stony Brook

INGRID M. PARKER
Department of Biology, University of California, Santa Cruz

GREGORY S. GILBERT
Department of Environmental Science, Policy, and Management,
University of California, Santa Cruz

The processes that control the transport, establishment, spread, and impact of invasive organisms underlie many of our concerns about sanitary and phytosanitary risk. In the United States and worldwide, our perception of risk has expanded from primarily economic and human health concerns to include risks to natural ecosystems. The effects of invasive non-indigenous species comprise one of the most apparent risks of globalization of international trade to both agricultural and natural ecosystems. Since Elton's (1958) first formal treatment of biological invasions as an ecological problem, ecologists have made great strides in understanding patterns of non-indigenous species introduction and establishment of self-sustaining populations (Williamson, 1996). Once a novel species is introduced to a region, the risk it poses depends on whether it will establish there and what impact it could have once established. There has been considerable progress in understanding which traits confer invasiveness (Rejmánek and Richardson, 1996; Reichard and Hamilton, 1997), but until very recently, much less attention has been given to developing rules for predicting, or even quantifying, impacts. Such rules could guide managers and policy makers in deciding which species can be safely introduced and which should be avoided, and help them set priorities to protect ecosystems.

In this paper, we review the impacts of invasions resulting from both planned and unplanned introductions from a biological perspective, pulling together examples and lessons learned from both agricultural science and the ecology of natural ecosystems. Although agricultural impacts and ecological impacts are rarely discussed in concert, we feel that the two are mutually illuminating; together they form a continuum that presents an array of challenges to managers and policy makers. First, we review the diversity of impacts and how they are measured from a strictly anthropocentric view, followed by a strictly ecological view, and then discuss what occurs when the two views are incongruent. In addition, we emphasize the utility of identifying trade and transportation pathways as a way to organize the diversity of potential invaders. Using vectors of introduction to define groups of invaders with similar biology and potentially similar impacts on natural systems may prove an efficient approach to identify targets for regulation and control. Often targeting the vector is easier than targeting each individual species.

We then focus on three aspects of predicting the risk of impacts from introductions of particular species: (1) predicting the successful establishment of invaders, (2) predicting the impact of species that establish themselves, and (3) assessing the uncertainty associated with these predictions. Progress in predicting establishment of certain invaders shows promise for informing policy on planned introductions, but relies heavily on detailed biological and geographical information about the species. Forecasting which invaders will then have the biggest impacts is even more challenging. Factors such as host range and dispersal ability can help gauge how big of a problem an invader will become. The prevalence of ecological idiosyncrasies, complex indirect effects and the possibility for synergistic effects among invaders, however, hinder our ability to predict ecological impacts. We finish with a discussion of prioritizing the management and control of invasive species based on impact, and the need for more consistent measures of impact.

To illustrate the main points that we emphasize in this paper, we have provided three detailed case histories of invasions. Through these case histories, we attempt to convey the importance that a biological perspective plays in assessing the degree of impact, as well as understanding the mechanisms behind that impact. We have intentionally chosen case histories that involve species native to the United States that have become problematic invaders in other parts of the world. This admittedly biased selection represents a suite of invaders less publicized within the United States that may feature prominently in current or future trade disputes.

IMPACT FROM AN ANTHROPOCENTRIC PERSPECTIVE

Impact from an anthropocentric view is often equated with the economic losses caused by a non-indigenous species or by the cost of its control or eradication. For example, the Office of Technology Assessment report on non-indigenous species lists as "high impact" those species that are significant pests of agriculture, rangelands, or forests and those that seriously foul waterways or power plants (U.S. Congress, 1993). Among several proposed measures of impact of non-indigenous species, Williamson (1998) quoted costs in pounds sterling for control of different weed species in the United Kingdom. Just as some environmentalists strive to attract wider public support for environmental protection measures by calculating the economic value of the services provided to humankind by biodiversity (e.g., Constanza et al., 1997; Daily, 1997), so have scientists raised public and political consciousness of the problem of non-indigenous species by pointing to the $100 million annual cost of fighting weeds in the United States (U.S. Congress, 1993) or the $400 million impact of zebra mussels in the Great Lakes over a five-year period (O'Neill, 1996).

Perhaps even more important to most people than their wallet is their health. Extensive research documents the negative impacts of non-indigenous pests and pathogens on human health, although, unfortunately, there is little sharing of information between the medical, public health, and epidemiological literatures on the one hand, and ecological and biological invasions literature on the other. Only occasionally does a medically oriented contribution appear in volumes devoted to invasion biology (e.g., Craig, 1993). Interestingly, invaders that cause human health hazards can have a psychological impact out of proportion to their real risks. A good example is the Africanized honey bee (*Apis mellifera scutellata*), a non-indigenous species made notorious by a relatively small number of deaths and immortalized by its nickname "killer bee."

In our review of the biological impacts of invasions, we include impacts on human health, on agricultural ecosystems, and on natural ecosystems. Throughout, however, we focus primarily on ecological impacts, both in natural and agricultural ecosystems, because here the diversity of interactions and the complexity of the issues best inform us of the range of risks we face and the sources of uncertainty in predicting consequences of biological invasions.

CASE STUDY 1: THE GRAPE ROOT LOUSE PHYLLOXERA—THE IMPORTANCE OF RECOGNIZING AND REGULATING VECTORS

The most devastating impacts of some invasive species have been economic and social, rather than environmental. Introduced agricultural pests, for example, cost millions in agricultural losses and control programs and have lasting effects on human populations. Such is the case of the grape root louse phylloxera (*Daktulospharaira vitifoliae*), an insect pest in the aphid family native to the Eastern United States. This pest first became problematic in its native range as European settlers in Eastern North America began to develop a wine-making industry in the seventeenth to nineteenth centuries. Viticulturalists discovered that European grape vines that had been imported because of their superior flavor (Morton, 1985) grew poorly in North American soils. In particular, the vines often shriveled and died within a few years of planting. Native American grapes, however, did not succumb to this affliction. The cause of the grape vine affliction was not discovered until it had been introduced into France and wiped out ancient vineyards. Exactly how phylloxera arrived in France is unknown. American vines had been introduced into France as early as the sixteenth century, but it was not until the advent of steamships allowing rapid crossings of the Atlantic that phylloxera could survive the trip (Stevenson, 1980). In the mid-1800s, colonists made several shipments of cuttings of American grapes to vine breeders in Southern France with the hopes of hybridizing the European species to the American. These vines were the probable vectors of phylloxera (Morton, 1985).

In 1863, centuries-old vineyards in the Rhone Valley began to show the effects of the phylloxera infestation. Within the next 10 years, not only had most of the vineyards in the Rhone Valley perished, but the infestation had spread throughout France, affecting more than 600,000 ha of vineyards (Pouget, 1990). By 1900, there remained only a few vineyards in all of France, and phylloxera had spread, presumably through the shipment of vine cuttings, to the rest of Europe (Oestreicher, 1996), Australia (Desdames, 1984; Buchanan, 1987), and South Africa (Oestreicher, 1996; Van Zyl, 1984). Later, phylloxera appeared in California, South America, New Zealand, and Japan.

The economic and social effects of the early European infestation of phylloxera were severe. Historical sources report the abandonment of entire communities in places like the Midi in Southern France where viticulture comprised the sole industry. The inhabitants migrated to Algeria, Argentina, or Chile, seeing little option for a livelihood in France or even Europe (Pouget, 1990). In fact, some historians have likened the social implications of the phylloxera epidemic in France to those of the potato blight in Ireland (Lukacs, 1996).

The French government responded to the crisis by appointing a special commission to study the pest and find a remedy. By 1887 they had discovered that grafting the European species to the rootstock of North American species provided a vine resistant to phylloxera but with the high-quality fruit of the European stock (Pouget, 1990). As the technology for grafting and selecting appropriate rootstock developed and farmers replanted, the wine industry slowly rebuilt.

Phylloxera, although still spreading in regions such as Australia and New Zealand (King and Buchanan, 1986), presents much less of a threat today because the cultural practice of grafting onto resistant roots has mitigated its impact. However, the more recent and disturbing discovery of a phylloxera strain in California that has overcome resistant rootstock threatens a resurgence of the phylloxera problem (Granett et al., 1985).

Phylloxera probably had very little direct ecological impact because of aspects of its biology. The host specificity of phylloxera and its probable mode of long-distance transport on grape vine cuttings, rather than autonomous flights, at least in some regions (King and Buchanan, 1986), confined the infestation to highly modified agricultural landscapes, some of which had been under cultivation for thousands of years. Had a more generalized pest been introduced or had phylloxera evolved the ability to use novel host plant species, its impacts may have extended to other agricultural systems or noncultivated ecosystems. To date, however, no research has looked for ecological impacts of the phylloxera invasion.

One unexpected negative effect of the phylloxera epidemic was the introduction of downy mildew into European vineyards, probably when

American rootstock was imported as the French sought to solve the phylloxera problem (Cowling, 1978; Lukacs, 1996). Thus the solution to one problematic invader served as the vector for introducing a second problematic invader—a story that has been repeated in other contexts as well (Simberloff and Stiling, 1996a).

The lessons to be learned from a biological perspective on the phylloxera invasion of the wine-producing regions of the world are threefold. First, to manage invasions we need to look for them and know what we are looking for. Although this idea may seem self-evident, it remains poorly implemented worldwide, partly because of a lack of well-trained taxonomists at points of inspection. The indispensability of well-trained taxonomists at all levels of the inspection service has a recent illustration in the discovery of Karnal bunt on wheat in the United States. Good taxonomic treatments coupled with modern diagnostic techniques led to a reliable program for detecting Karnal bunt and differentiating it from similar fungi. This program has saved the $5 billion wheat export market from international prohibition of cereal and grass seed from the United States (Palm, 1999). Second, we recognize that ecological and evolutionary interactions between pests and their hosts play an important role in phytosanitary risk, as they do in the health risks of emerging human diseases (Ewald, 1994). Third, our strategies for managing the impacts of an invasion often carry with them their own invasion risks, such as the downy mildew associated with American grape vines. All three of these lessons can be understood in the context of vectors, or routes of introduction, in ways that increase our ability to predict potential invasions.

VECTORS

By examining routes of introduction of invaders, planned or unplanned, we find patterns that may help us predict the likelihood of particular kinds of impacts associated with different kinds of introductions. In many cases, regulating the vectors may prove easier than regulating the organisms themselves. A vector-based perspective can help organize the complexity of invasion impacts into approachable subunits, thereby potentially providing useful generalizations to guide policy decisions. Specifically, analysis of vectors offers two advantages: (1) identifying a restricted group of taxa likely to have similar impacts that can be managed in a particular way and (2) using the vector to manage the risks of a particular species.

Identifying Patterns of Species Introductions and Impacts

One of the most well-known examples of a vector for introducing non-indigenous organisms is ballast water. Ballast water taken on by ships in one part of the world, then expelled in another region, can efficiently transport a diversity of marine and freshwater organisms including bivalves, crustaceans, and fish (Mills et al., 1993; Carlton and Gellar, 1993; Cohen and Carlton, 1998). Locke et al.(1993) estimated that 800 million liters of ballast water are dumped into the North American Great Lakes each year. The impacts of the organisms that have established through such means can include mechanisms generally important in all ecosystems, such as competition and predation, as well as mechanisms specific to that suite of organisms, such as the voracious filter feeding of the Asian clam (*Potamocorbula amurensis*) (Werner and Hollibaugh, 1993) or the fouling of industrial pipes by the invasive zebra mussel (*Dreissena polymorpha*) in the Great Lakes (MacIsaac, 1996). Ballast water is a good example of a vector targeted for regulation, as the zebra mussel disaster resulted in the passage of federal regulations requiring the exchange of fresh water ballast for salt water ballast before ships enter the Great Lakes (U.S. Coast Guard, 1993). These regulations, in theory, should greatly reduce the risk of invasion. Nevertheless, it comes as little surprise that several prolific invertebrates have invaded the Great Lakes since the implementation of preventative measures (Ricciardi and MacIsaac, 2000). Lack of complete compliance with ballast exchange policy (Locke et al. 1993), and the possibility that nonindigenous organisms may be transported on the hull or other parts of ships, suggest that we may need to reevaluate the implementation and enforcement of the policy periodically.

The most obvious vector for a pest or pathogen is its own host species, and consequently many phytosanitary regulations have focused on this pathway. Perhaps surprisingly, hosts as vectors may take menacingly varied forms. For example, New Zealand officials intercepted the fungus *Bipolaris maydis*, which caused the epidemic of Southern corn leaf blight that devastated the United States corn production in 1970, in packages of popcorn imported from the United States (Scott, 1971). We generally expect new pests and pathogens to come from the area of origin of the host species (Thomas, 1973). Because the majority of important crop species originated outside of North America, the United States may seem an unlikely origin of pests in this regard. However, the U.S. Department of Agriculture (USDA) Germplasm Resource Information Network (GRIN) lists 418 species or subspecies of flowering plants native to the United States with a recognized economic importance for food, construction, fuel, or forage (see Table 5-1).

TABLE 5-1. Counts of Plant Species Native to the United States That Have a Known Economic Importance

Economic Importance	Number of Species
Food	105
Construction	206
Fuel	28
Forage	126
Weeds	284
Alternate disease host	14

Source: USDA Agricultural Research Service 1999.

It is important to note that pests and pathogens commonly show rapid evolution (Ebert, 1998; Thomson, 1998), and host shifts can occur unexpectedly (Strong 1984) leading to disease or infestations of introduced crops by pathogens from the new range of the crop. Novel pathogens can then be transported back to the region of origin of the crop, or to other regions through unexpected vectors. The fire-blight pathogen (*Erwinia amylovora*) in New York illustrates such a transfer. This pathogen shifted from its previously known host, wild crabapple (*Crataegus* sp.), to cultivated apples, which had been introduced from Eastern Europe. Following the subsequent introduction of the pathogen to England, it spread throughout Europe, causing extensive losses in the apple industry (Van der Zwet and Beer, 1992). Similarly, the introduction of North American rainbow trout (*Onchorynchus mykiss*) to Europe for trout farming was followed by an apparent host switch of the Eurasian fish parasite (*Myxobolus cerebralis*) that causes whirling disease (Hedrick et al., 1998). This parasite was then transported to North America via infected live or frozen trout, where it imperils both commercial rearing operations and natural populations (Hoffman, 1970). Host switching and rapid evolution of pests and pathogens, coupled with global distributions of many crops and noxious invasive species, complicates our ability to predict routes of introduction and invasion.

Cargo and luggage also provide passage for a wide diversity of "hitchhikers" that may represent serious invasion threats. Similarly, some kinds of agricultural or horticultural introductions (e.g., rootballs packed with soil) are of particular concern not because of the plants themselves, but because they may harbor a diversity of unexpected organisms. Of the potential pest organisms that the Australian Quarantine Inspection Service (AQIS) intercepted on entry to Australia from the United States between 1990 and 1998, 80 percent were found in shipping containers or sea freight, and an additional 15 percent were found in air freight (AQIS, 1999). Intercepted organisms ranged from nematodes to frogs, associated with everything from pottery to logs. Of 8,243 interceptions during that period, 42 percent were encountered on fruits (60 percent mites, 10 percent hemipteran bugs, and 10 percent thrips) and 37 percent on timber (92 percent beetles).

Using Vectors to Regulate Specific Risks

A second use of vectors as an approach to understanding and regulating invasions involves the management of particular risks associated with a particular species. For example, brown tree snakes were introduced into Guam from New Guinea or Australia as stowaways in shipments of derelict war equipment shortly after World War II (Fritts and Rodda, 1998). By the 1980s, these predators had extirpated or drastically reduced population sizes of most of the 10 species of native birds, as well as reduced the populations of native fruit bats and lizards (Fritts and Rodda, 1998; Savidge, 1987). Brown tree snakes also have had major economic impacts on poultry operations and caused power outages by climbing on transformers (Fritts and Rodda, 1998). Now, Hawaii and other Pacific islands are eyeing the problems on Guam with alarm. Flights and cargo from Guam into Hawaii undergo mandatory inspection to prevent the introduction of the brown tree snakes. As a tribute to their success, Hawaii remains free of the snake, although six snakes were intercepted between 1981 and 1994 on incoming airplanes (Fritts, 1999).

Another cargo hitchhiker most effectively regulated through its vector is the Asian long-horn beetle (*Anoplophora glabripennis*). Beetle larvae burrow in untreated wood used in packing crates in shipments from China and have emerged as adult beetles in several U.S. cities. The mobile beetles are voracious pests of a number of important tree species and have caused widespread death of trees in New York and elsewhere (Haack et al., 1997).

IMPACT FROM AN ECOLOGICAL PERSPECTIVE

Measuring the Ecological Impact of Invaders

A biologist trying to determine the impact of an invader on native species and ecosystems faces a more difficult challenge than an economist evaluating that invader in financial ("pest") terms. There are many different approaches to measuring impacts, and there appears to be little agreement among ecologists over how impacts should be quantified and compared among invading taxa (Parker et al., 1999).

The simplest measure of impact is the area of land occupied by an invader (Dombeck, 1996; Schmitz et al., 1997). Using range to represent impact, however, assumes that all invaders have effects of a similar magnitude on local biological communities, which is clearly not the case (Williamson, 1996; Wonham et al.,2000). A complete assessment of impact would incorporate the range of the invader, its abundance, and its local effects (Parker et al., 1999). However, the local effects depend on the ecological interactions between an invader and its host community or ecosystem (Drake, 1983). Determining local effects of an invader, therefore, represents the greatest challenge to predicting the impacts of a particular invader in an ecosystem. Local impacts can be measured at five scales of ecological organization: traits of individuals, genetic characteristics of populations, abundance and dynamics of populations (within species), communities of multiple species, and ecosystem processes.

Individuals

Invaders can have a variety of impacts on the characteristics of individuals. For example, invaders can compete with natives causing poor growth and reduced individual size (Gentle and Duggin, 1997) or altered morphology, such as rooting depth (D'Antonio and Mahall, 1991). Invaders can also cause changes in behavior of native animals, such as invading predacious fish that alter the habitat use and diet of native fish in rivers (Brown and Moyle, 1991). Impacts on individuals of a species can have important implications for the viability or dynamics of the whole population as well, but can be easier to measure than population parameters.

Genetics

Sometimes non-indigenous species are introduced to areas already inhabited by closely related native species. If these native-non-indigenous pairs of species can interbreed, then genetic interchange between two species can lead to a loss of the unique genetic makeup of the species (Echelle and Conner, 1989; Rhymer and Simberloff, 1996). Such hybridization and introgression can lead to a virtual extinction of native species through "genetic pollution," especially when the invader becomes much more common than the native species. In fact, of 24 federally listed species in the United States that have become extinct since the enactment of the Endangered Species Act, three have done so through hybridization with non-indigenous species (McMillan and Wilcove, 1994).

Populations and Species

Species-level measures of impact are often used in ecology and conservation. Extinction of native species is arguably the most dramatic impact of invasive species. Recall, for example, the brown tree snake that caused the extinction, or near extinction, of most of the native bird species of Guam (Savidge, 1987). Extinction of native species, however, characterizes relatively few invasions (Simberloff, 1981). Reduced population sizes or local extinctions appear more common, but changes in population sizes of native species after invasion by a non-native can vary greatly in magnitude and even direction. Researchers documented this wide variation with changes in abundance of terrestrial arthropod species before and after invasion of the fire ant (*Solenopsis invicta*) in Hawaii (see Figure 5-1). Although most species showed reduced abundance because of competition or predation by fire ants, other species, such as the scarab beetle *(Myrmecaphodius vaticollis)* increased in abundance (Porter and Savignano, 1990).

Communities

Community measures, such as changes in species diversity or richness, incorporate effects on many individual species, yet provide a simple and interpretable summary of impact. Although these measures do not reflect the magnitude of change in any one species, they do provide a sense of how much an invader interacts with community members. Community measures have the further advantage that they are potentially comparable across communities with different species assemblages.

96

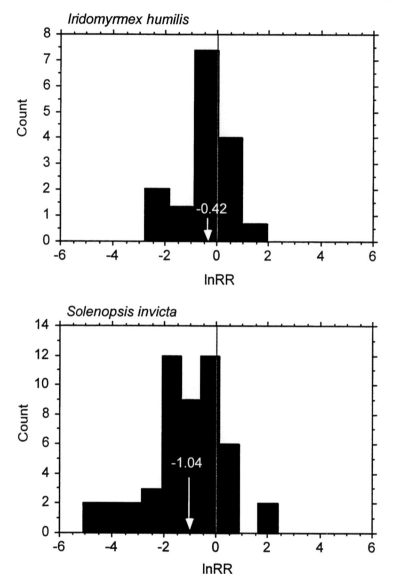

FIGURE 5-1. Frequency Histogram of the log Response Ratio (lnRR) for Ant Impacts. lnRR is a measure of effect size used in meta-analysis(Hedges, 1996), here defined as lnRR = ln(abundance with the invader/abundance without the invader). Each bar represents the number of sampled resident invertebrate species that responded to the invasion by one of two introduced ant species at each effect size category. When lnRR = 0, there was no change in density of the native species, when lnRR < 0 the native species density declined in the presence of the invader, and when lnRR > 0 the resident species increased in density in the presence of the invader. Arrows and numbers in white show means. Data on *Iridomyrmex humilis* were taken from Cole et al., (1992); data on *Solenopsis invicta* were taken from Porter and Savignano (1990).

Ecosystems

Invaders that affect ecosystem processes such as nutrient cycling or disturbance regimes are viewed by some ecologists as the most problematic (Vitousek and Walker, 1989; Mack and D'Antonio, 1998). These biologists reason that changing ecosystem processes "changes the rules of the game" in a way that influences many, if not all, of the component species. For example, *Myrica faya*, an introduced nitrogen-fixing shrub in Hawaii, colonizes the nutrient-poor soils of recent lava flows and increases soil nitrogen levels (Vitousek and Walker, 1989). These changes in the nutrient cycle are thought to affect patterns of succession on the lava flows (Vitousek et al., 1987). The large amount of water used by the invasive saltcedar trees (*Tamarix ramossissima*) (Carman and Brotherson, 1982) lowers the water table in ephemeral or permanent wetlands in North American arid zones and can eliminate habitat for native migratory birds (Neill, 1983).

CASE STUDY 2: THE MOSQUITO FISH—WHEN ANTHROPOCENTRIC AND ECOLOGICAL PERSPECTIVES CLASH

The ecological impacts of some biological invasions contrast with their economic and social impacts. This situation characterizes some invaders that also serve as biological control agents. Resolutions of such conflicts require policymakers to carefully weigh very different sorts of impacts.

From a purely ecological perspective, the mosquito fish (*Gambusia affinis affinis* and *Gambusia affinis holbrooki*, two subspecies) exemplifies a biological control agent gone awry. After the discoveries in 1898 that mosquitoes transmit malarial parasites to humans, and in 1900 that a species of mosquito carries yellow fever, interest in mosquito control grew (Krumholz, 1948). The mosquito fish, native to the southeastern United States, quickly rose as a popular control measure. True to their name, mosquito fish consume large numbers of aquatic mosquito larvae. The American Red Cross, the International Health Board, and the Rockefeller Foundation jointly mounted a program to control mosquitoes by introducing mosquito fish to many regions of the world, including parts of the United States where the fish was not native (see Figure 5-2). This program gave little thought to potential negative ecological effects (Lloyd et al., 1986).

Some mosquito fish introductions studied not only effectively controlled mosquito populations (Gerberich, 1946; Krumholz, 1948), but reduced incidence of malaria (Howard, 1922; Gerberich, 1946), and in one case even increased the rate of human population growth (Holland, 1933). Mosquito fish proved prolific and tolerant of a wide range of environmental conditions, including pollution (Krumholz, 1948; McKay, 1984; Lloyd et al., 1986). In retrospect, it comes as little surprise that some of the same characteristics that made the mosquito fish a successful biocontrol agent also made it a successful

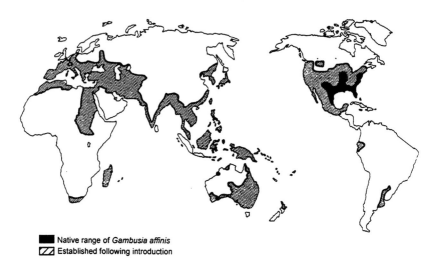

■ Native range of *Gambusia affinis*
▨ Established following introduction

FIGURE 5-2. Worldwide Distribution of *Gambusia affinis*. Modified from Lloyd et al. (1986) showing that *G. affinis* now enjoys a strikingly broader distribution than its native distribution. Shaded regions representing establishment following introduction indicate that *G. affinis* occupies suitable habitats in those regions. Reproduced by permission of John Wiley & Sons, Australia.

invader. Their broad environmental tolerances allowed them to thrive in many regions of the world. Their rapid population growth and large appetite put them in competition with native surface-feeding fish. In addition, mosquito fish consume eggs and young of native fish and amphibians. Australian researchers, in particular, have conducted detailed research on mosquito fish impacts, carefully documenting declines in small surface-inhabiting native fishes, as well as juveniles of important game and food fish species (McKay, 1984; Howe et al., 1997). Mosquito fish have been implicated in the decline of 35 fish species throughout the world (Lloyd, 1989), as well as three Australian frog species (Webb and Joss, 1997). At least one of these frogs, *Litoria aurea*, is listed as threatened (Morgan and Buttemer, 1996). Experimental research on the impact of mosquito fish suggests that it could have ecosystem impacts by consuming aquatic invertebrates and thus altering the trophic balance of still-water systems. Ultimately these changes could lead to accumulation of nutrients, increased biotic growth, and oxygen-deficient water (i.e., eutrophication; Hurlburt et al., 1972).

To evaluate whether the net consequence of introducing mosquito fish turned out, on balance, positive or negative, one must weigh the ecological effects against the human health effects. To put the negative effects in perspective, it is instructive to compare the ecological impacts of using mosquito fish to control mosquitoes with those of alternative methods. Previous methods were either less effective and more expensive, such as applying oil films to the surface of still water (De Buen, 1929, in Gerberich, 1946), or imposed a different suite of ecological and human health risks, such as the application of DDT (Krumholz, 1948). In comparison to the impact of other control measures, the impact of mosquito fish may seem less egregious. Fish biologists in Australia, however, contend that several native Australian species could serve in mosquito control, obviating the need to introduce non-native species (Lloyd et al., 1986). In this case, knowledge of ecological impacts combined with a good understanding of the biology of the system could motivate more appropriate and safer alternatives that were overlooked or unavailable in the past. Even with all of the available information on the negative ecological impacts of the mosquito fish, however, policymakers may still decide to introduce the fish into new areas. Although ecological information may not always drive policy, it is critical to "have all the cards on the table" for society to make informed decisions.

The mosquito fish represents just one of many invaders for which we will have to develop regulations that weigh the relative costs and benefits to society. Of the 284 species of important international weeds that are native to the United States, GRIN (USDA, 1999) lists 39 species that have both weed potential and important economic uses. This duality characterizes the well-known case of the Eurasion honey bee, *Apis mellifera*, which was brought to North America by European colonists for honey and wax production. Apiculture has since burgeoned into a $10-billion industry, playing a key role in the pollination of many insect-pollinated crop plants (Robinson et al, 1989). However, ecologists have documented negative effects of honey bees on native pollinators and plants, through competition for floral resources and inadequate pollination (Roubik, 1982; Kato et al., 1999; Goodell, 1998; Paton, 1997).

PREDICTING OUTCOMES OF SPECIES INTRODUCTIONS

When we consider both economic and ecological impacts, the uncertainty involved in predicting the impacts of unplanned introductions can be overwhelming, even if we employ a vector-based approach to focus on a subset of potential invaders with similar impacts (e.g., ballast water introductions). One attempt to quantify risk at the level of a vector was an assessment of the potential cost posed by untreated Siberian timber to North American forest ecosystems and the U.S. timber industry (USDA, 1991; Ruesink et al., 1995). This assessment suggested that losses due to unintentional introduction of organisms associated with larch could have reached $58 billion (USDA, 1991).

Planned introductions, for which we at least have an identified organism, may provide a better starting point to test our ability to predict the impacts of species introductions. Intentionally introduced organisms offer us a variety of appealing traits—an attractive shrub, an herbivorous insect that controls a noxious weed, or an affectionate pet. Yet these desirable organisms may sometimes themselves become invasive pests. We discuss three components important to assessing the risk of a planned introduction: (1) predicting which species will successfully establish in a new region, (2) predicting which of those species will have large impacts, and (3) assessing the degree of uncertainty involved in the prediction.

Quantitative and Experimental Approaches to Identifying Potential Invaders

Predicting the outcome of species introductions is an area of active research in ecology. So far, few general rules of thumb regarding which introduced species will establish self-sustaining populations apply to all kinds of non-indigenous organisms. For some taxonomic groups or ecologically similar groups of species, however, biologists have identified characteristics that fairly consistently correlate with invasiveness. The methods biologists use to reach these conclusions typically involve scoring introductions of known outcome as invasive or not invasive, then looking for correlations between invasiveness and various attributes of the introduced species, such as life history traits or biogeography, using multivariate statistics (e.g., Bergelson and Crawley, 1989; Perrins et al., 1992; reviewed in Ruesink et al., 1995).

At the taxonomically broadest end of the spectrum, Reichard and Hamilton (1997) developed a decision analysis to evaluate which characteristics were associated with invasiveness of woody plant species that had been introduced into the United States. Starting with a multivariate approach that included many different traits, they found a strong tendency for woody plant species that were invasive in other parts of the world also to be invasive in North America. Some attributes proved more useful for regional models than for the continental-scale model. For example, having an Asian origin indicated non-invasiveness in the large-scale model but invasiveness in a regional model using data from the southeastern United States. Their results suggest that models incorporating geographic attributes may apply only to the region for which they were developed and may not be generalizable to other geographic regions.

Reichard and Hamilton (1997) then used these distinguishing attributes of invaders to construct a "decision tree" for use in deciding which woody tree species should be introduced and which should not. The decision tree resembles a flow chart and presents a series of questions regarding the presence of particular traits, starting with those most strongly associated with invasiveness. The answers to these yes/no questions form dichotomous branches leading to either a question about the next most important trait or a recommendation to accept, reject, or further study the species proposed for introduction. Their decision tree relies on information about the introduced species that is relatively easily obtained from the literature or herbarium records, which makes it a practical tool for managers. In validating their decision tree using all woody plant introductions, invasive and non-invasive, they correctly rejected 88 percent of the pest species and unconditionally accepted for admission only 7 percent of the invasive species. Predictive power for non-invaders was lower with 46 percent admitted unconditionally, 18 percent rejected, and 36 percent recommended for further analysis. The decision analysis approach offers an efficient and flexible screening process for proposed, planned introductions.

Because traits associated with invasiveness often differ among taxa, the greatest predictive power may lie within narrow taxonomic groups. Rejmánek and Richardson (1996) examined pine trees (*Pinus* sp.) for traits predictably associated with invasiveness. They used multivariate statistical techniques to distinguish between invasive and non-invasive pine species on the basis of ecological and life history characteristics. According to their model, invasive pines share characteristics associated with long-distance dispersal and rapid individual and population growth rates (e.g., small seeds, short minimum juvenile period, and a short time interval between large seed crops). The results appeal to our ecological sensibilities because they underscore traits that influence processes of species establishment and spread, as well as formation of persistent, self-sustaining populations. Exceptions to the pattern, such as *Pinus pinea,* a large seeded, vertebrate-dispersed tree species, also had relatively accessible ecological explanations. Despite its more specialized mode of dispersal, this species has become moderately invasive in regions of the endemic South African fynbos ecosystem where introduced squirrels (*Sciurus carolinensis)* disperse its seed (Richardson et al., 1990).

In an attempt to generalize from a taxonomically narrow model, Richardson et al. (1990) used the life history criteria that distinguish invasive and non-invasive pine species to categorize *Banksia* spp. into functional (ecologically similar) groups. Some of these native Western Australian trees and shrubs have begun to arrive in South Africa and share similar suites of attributes with other shrubs invasive in the African fynbos. Functional groups that possess traits associated with invasive pines are identified as high risk, but the authors point out that belonging to a functional group does not necessarily guarantee success or failure of an invasion. As in the pine and squirrel case, idiosyncrasies of the invader or its interactions with the recipient community often play an important role in success or failure. Of course, we might successfully predict some of those idiosyncrasies if we knew the biology of the system well enough. For example, some *Banksia* identified as high risk are susceptible to a pathogenic fungus already present in South Africa, which may inhibit potential invasions (Richardson et al., 1990). Richardson et al. (1990) took advantage of the large body of knowledge about these two groups generated primarily by their economic importance. Clearly, a sound understanding of the basic biology, ecology, and natural history of any potential invader and recipient community is requisite for making accurate predictions. Sadly, we often lack this seemingly basic information.

Some researchers have tried to gather some of the missing ecological information needed for predictive models. Forcella et al. (1986) used biogeography and empirical ecophysiology combined with multivariate statistics to examine the relationship between species characteristics and invasiveness in *Echium,* a genus of herbaceous and shrubby plants. Species known to be invasive in Australia had broad native distributions, which may reflect their probability of introduction. They also showed rapid seed germination under a range of conditions, which may reflect their probability of establishment. Their model could serve to evaluate species of *Echium* of unknown invasiveness, although the empirical studies needed for each unknown species would require substantial time and funding.

The results of the above predictive models highlight the importance of good ecological, biogeographical, and historical data. These models can best serve managers and policymakers when the information required about the species in question can be obtained relatively easily. For instance, the woody tree species example requires information about the invasiveness of the species in question in other parts of the world. Biogeography of *Echium* proved useful in the model by Forcella et al. (1986). This type of information is easily amenable to Internet-accessible databases. The further development of such predictive models will depend on the creation, maintenance, and accessibility of large databases of invasive species worldwide.

Experimental plantings of individual species into areas beyond their native range potentially offer an alternative to predictive models and show promise in determining if a particular species will become invasive in a region. Experimental plantings could have application in fields such as agriculture and horticulture, in which planned introductions are the rule. This approach may prove especially powerful if combined with manipulative studies to test the success of introductions over a variety of environmental conditions (Mack, 1996). The drawback of conducting experimental transplants of nonindigenous plants or animals outside of their native range is the risk of escape. The unfortunate consequences of this risk are manifested in the escape of the gypsy moth and the Africanized honey bee, both of which were brought to the Americas to investigate their cultivation for commercial purposes. In making use of experimental introductions, researchers must take extreme precautions to prevent escapes (including the escape of genes from introduced plants, Ellstrand and Hoffman, 1990), and the implementation of these experiments should be regulated by agencies designed to monitor nonindigenous species introductions. Conducted properly, these experiments will be costly, but could provide the desired predictive power for assessing risk of introductions

Predicting the Impacts of Those Invaders

Predicting invasion success is a necessary requirement for predicting impact, but it is not sufficient. Within groups of successful invaders, only a proportion will have a large impact (Williamson, 1996). Defining impact, of course, is essential to evaluating it (Parker et al., 1999). When measuring impact of an invader, the parameters chosen can affect greatly the magnitude of the impact detected. To illustrate this point we refer again to ant invasions, specifically two studies on the invasive ant species *Iridomyrmex humilis* (Cole et al., 1992) and *Solenopsis invicta* (Porter and Savignano, 1990); in each case the change in abundance associated with invasion was measured for a large collection of resident invertebrates species. For both studies, the size and even direction of the effect varied greatly among response species for both invaders (see Figure 5-1).

Lessons about predicting impact for introduced species come from the regulatory process for evaluating potential biological control agents, that is, non-indigenous species introduced to control a specific weed or pest (DeBach and Rosen, 1991). Before introduction, researchers screen potential biocontrol agents to see if they are likely to survive and produce viable populations in the new range (i.e., invasion success; Waage and Greathead, 1988). Although these tests are themselves time and resource intensive, a further step would be to evaluate whether sustained populations of the control agent would really inflict a significant demographic impact on the weed or pest. Currently, researchers rarely consider this last step before releasing biological control agents (McEvoy and Coombs, 1999). The process for evaluating or predicting impact seems complex for biological control agents, yet with other non-indigenous species the process becomes even more complex because we have to consider impacts on many more than just one target species, as well as community or ecosystem effects.

Ecologists have made little progress in predicting which invaders will have a big impact (Parker et al., 1999). One early idea proposed by Darwin (1859) and Gause (1934) suggests that ecologically similar species will interfere with each other more strongly than ecologically different species. Therefore, if an introduced species successfully establishes, it might have the biggest impact on species that are most closely related to it if related species play similar ecological roles. A recent meta-analysis (Goodell et al., 2000) combined results from seven studies showing responses of 61 resident species to a variety of insect invasions and found support for the idea that invaders have larger competitive impacts on confamilial resident species than on more distantly related species (see Figure 5-3). The "close relatives" generalization aids us less in predicting which non-indigenous species will have big impacts overall, and more in predicting which species may suffer the largest impacts by an invader. From a practical perspective, this generalization may contribute most to protecting particular suites of rare or sensitive native species from specific introductions.

Another fairly well-accepted idea in invasion biology is that introduced species have the greatest impacts when the invader performs a novel function in the recipient community (Elton, 1958; Simberloff, 1991). The idea is that these novel types of species can dramatically change the ecological context for many species at once. Because the "close relatives" generalization specifically applies to competition between ecologically similar pairs of species, it does not necessarily conflict with the "novel function" generalization. Predators on oceanic islands (Elton, 1958) and nitrogen-fixing shrubs in nutrient-poor systems (Vitousek and Walker, 1989) comprise some classic examples of species that perform novel functions. A related idea suggests that species that dramatically change disturbance regimes (e.g., frequency of fires or floods) impose very large ecosystem-level impacts (Mack and D'Antonio, 1998) and will also have large impacts on all the component species in a system.

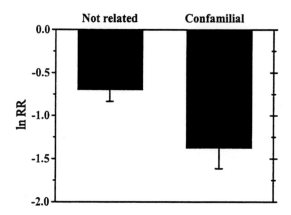

FIGURE 5-3. Mean Effect Sizes (±1 SE) of Insect Invaders on Resident Confamilial Species (n=27) Versus More Distantly Related Species (n=34). Data compiled from seven studies of impact and represented as log response ratios (see Figure 5-1 legend). These data only consider competitive interactions. Note that smaller (more negative) numbers indicate that invaders were associated with larger reductions in population abundances.

These generalizations are interesting and represent an important first step in defining what is a large impact, but they guide prediction only in a very limited context. That is, identifying the potential change in population dynamics of a target host plant, identifying the species ecologically most similar to the invader, or identifying novel functions can be done only in the context of a particular natural community. In fact, many have argued that we should only study invasions as an interaction between the invader and the host community (Drake, 1983; Simberloff, 1986). In light of the great diversity of ecological communities within the boundaries of the United States alone, matching potential introductions with native communities seems daunting indeed.

Before we give up hope on producing generalizations about which species tend to have large impacts, it is important to realize that invasion biology is an extremely young science. So far, there have been almost no attempts to synthesize information because, until recently, we did not have very extensive data on the impacts of different invaders. Scanty published information even for the most infamous non-indigenous species (Hager and McCoy, 1998) serves as a reminder that many basic questions remain unanswered. The poverty of our knowledge of the biology and natural distributions of insects and pathogens adds another dimension of difficulty to predicting which invaders will inflict the most damage.

CASE STUDY 3: THE CRAYFISH PLAGUE AND THE SIGNAL CRAYFISH—LIMITS TO PREDICTION WHEN SPECIES INTERACT SYNERGISTICALLY

Crayfish plague is a disease of crayfish native to the United States. This oomycete pathogen (*Aphanomyces astaci*) was first reported in Europe in 1860 where it devastated populations of the native Noble crayfish (*Astacus astacus*) in Northern Italy (Laurent, 1997). The exact mode of introduction is unknown, but crayfish plague most likely crossed the Atlantic in fresh ballast water. The disease subsequently spread throughout the eastern parts of Europe, Russia, and the Baltic states from the 1870s through the 1920s (Alderman, 1996). The mode of spread likely included human transport of crayfish for trade and aquaculture within Europe, transport of infected fishing equipment (Unestam, 1973), and possibly mammalian and bird predators of crayfish that move between isolated bodies of water (Taugbol and Skurdal, 1993).

Native European crayfish play an important ecological role in freshwater ecosystems. Crayfish are thought to keep aquatic plant growth in check and their absence is associated with overgrowth of lakes (Unestam, 1973). At one time, several species also were important in fisheries. All five species of native European crayfish show extreme susceptibility to crayfish plague. As the crayfish plague spread through Europe, reduced population sizes and local extinctions of native species occurred, causing a marked decline in overall crayfish densities. This decline undoubtedly had ecological effects in addition to collapsing the crayfishing industry (Unestam, 1973; Nylund and Westman, 1992).

In the mid-1960s a new phase of commercial interest in crayfish bloomed in Europe, which included the introduction of the Signal crayfish (*Pacifasticus leniusculus*) native to the western United States. The large size and overall hardiness of the Signal crayfish made it an attractive commercial species. In addition, its natural resistance to the crayfish plague pathogen offered it as an ideal replacement for the native species in the plague-ravaged waters of Europe.

Although importing Signal crayfish has undoubtedly bolstered the commercial crayfish industry, it has brought a double dose of harmful effects on native species. The most damaging impact of the Signal crayfish has been its role as a vector of crayfish plague. These introductions brought with them several new infestations of crayfish plague. Transport of the Signal crayfish among European countries has facilitated spread of the pathogen. DNA evidence has linked recent range extensions of crayfish plague into Spain in 1965 (Diéguez-Uribeondo et al., 1997) and Britain in the 1980s (Lilley et al., 1997) to these new introductions, although other new outbreaks are thought to stem from recent Signal crayfish introductions as well (e.g., Greece in 1982; Lowery and Holdich, 1988).

Increased spread of crayfish plague via Signal crayfish introductions appears to have increased the frequency of infections and also may have prevented the extinction of the pathogen with its dying native hosts in isolated bodies of water. This pattern of extinction has been observed in Ireland, where alien crayfish have not been imported (Matthews and Reynolds, 1992). Beyond its role as a disease vector, Signal crayfish also directly affects aquatic organisms in the areas where it has been introduced. It is larger and more aggressive than its European counterparts and can have a larger impact on its competitors and prey. These attributes have reduced densities of native fish, as well as further reduced native crayfish densities. For example, Signal crayfish both eats and competes for shelter with native British fishes: the stone loach and the bullhead. Together these effects cause a negative relationship between crayfish density and the density of these stream fish (Guan and Wiles, 1996). In a Swedish lake, Signal crayfish outcompeted native Noble crayfish for shelter, leading to greater perch predation on the native species (Söderbäck, 1994).

The evidence suggests that the crayfish plague and the introduced Signal crayfish have synergistic impacts. This phenomenon was named "invasional meltdown" in a recent paper by Simberloff and Von Holle (1999). We do not know how generally this pattern applies to invasions, but it certainly represents a type of unexpected effect that could foil attempts to predict impact of an invasive species.

Indirect Effects of Invaders May Limit Predictability

Like the Signal crayfish and its impact on natives through shared pathogens, species can have far-reaching indirect effects mediated through a complex of ecological interactions. For example, a seldom recognized threat of introduced plant diseases comes from alternate hosts (i.e., obligate second host in a pathogen life cycle) or alternative hosts (i.e., additional host to a pathogen of a crop species). GRIN lists 14 plant exports native to the United States in this category (Table 5-1; USDA, 1999). Of particular interest are plants of potential horticultural value that serve as alternate hosts for heteroecious rust fungi that attack other economically important plants. In such cases, the rust fungi must pass alternately through two host species to complete their life cycle. Introduction of the alternate host could potentially introduce the pathogen as a hitchhiker. Even if the pathogen is not introduced on the alternate host, the presence of the alternate host creates the opportunity for disease development should the pathogen be introduced independently at some time in the future. If strong indirect effects of invaders are common, efforts to predict the impact of invasive species will be particularly difficult (Simberloff and Stiling, 1996b).

Assessing the Degree of Uncertainty

Because biological systems are complex and never fully understood, predictions of risk will always carry with them some uncertainty. As shown by the above examples, indirect effects can be extremely difficult to forecast a priori without detailed information about both the organism and the recipient ecosystem. We believe that certain types of invasions are more likely than others to have indirect effects, or highly unpredictable effects, and that explicitly identifying this degree of uncertainty forms an important step in risk assessment. We do not suggest that more predictable invasions necessarily will have smaller impacts, only that the impacts will be easier to project and possibly quantify.

Organisms with the highest degree of predictability in terms of impact include highly host-specific pathogens or pests in agroecosystems. For host-specific, seedborne pathogens, the introduced pathogens will not spread beyond the field planted with the seeds; thus their impacts represent a predictable quality control issue with minimal economic or ecological importance. Similarly, we also expect to have reasonably high predictive power for impacts of host-specific pests and pathogens that cause epidemics in previously unaffected geographic areas. Many problematic agricultural invasions fall into this category, such as zonate leaf spot of sorghum (caused by the fungus *Gloeocercospora sorghi*), which arrived in Venezuela on seed sent by the USDA for experimental purposes and subsequently became widespread (Ciccarone, 1949). In cases such as these, as well as for releases of biological control agents, the final impact on a host can vary in different areas and under different ecological conditions. In addition, as we have discussed, it is difficult to eliminate the possibility of a future switch in host use. Nevertheless, their impact should follow more predictable patterns than that of pests, pathogens, or other introduced organisms with multiple hosts.

Of intermediate predictability are impacts of host-specific pests or pathogens (or parasites or predators) whose host species is a dominant species or "keystone species" (Power et al., 1996) in the invaded natural ecosystems. The Chestnut blight pathogen (*Cryphonectria parasitica*), introduced to the United States on European logs, led the American chestnut to the brink of extinction and changed the dominant tree over millions of hectares of forest (Anagnostakis, 1987; von Broembsen, 1989; Jarosz and Davelos, 1995). Also of intermediate predictability are impacts of species that attack not just one but rather a fairly restricted suite of hosts, such as biological control agents with undesirable effects on nontarget species (Louda et al., 1997; Louda, 1998). We may not predict their impacts perfectly, but with good basic information about the likelihood of alternate host use, we should make reasonable assessments of the risks.

Among introduced species that likely affect many resident species, our best chance of predicting impacts lies in those invaders with restricted habitat use. For example, many introduced organisms will persist only in highly disturbed agroecosystems. Similarly, we might anticipate fairly easily the impact of invasive aquatic plants such as the water hyacinth (*Eichhornia crassipes*) because of their restricted habitat use and consequently restricted set of interacting species (Schmitz et al., 1997).

In contrast, the species with the most difficult impacts to assess, and therefore those with the biggest uncertainty, are species that can invade a variety of natural ecosystems and that interact with many different native species. The oomycete pathogen *Phytophthora cinnamomi* has had a devastating impact on the eucalyptus forests in Western Australia, where the pathogen attacked more than a third of the species present in the forest (Shearer and Dillon, 1995), causing a wholesale reshuffling of the ecological interactions within that plant community. Opportunistic predators such as feral house cats eat small mammals, songbirds, reptiles, and insects in proportion to their availability in the environment (Pearre and Maass, 1998). Where introduced or domesticated animals are abundant, they comprise the bulk of the cats' diet (Langham, 1990), but in other areas cats may be important predators of endemic animals, such as the endemic lizard, *Urosaurus auriculatus,* on Socorro Island, Mexico (Arnaud et al., 1993). The highly variable impact of an organism like the house cat, then, will hinge on characteristics of the particular area into which it is introduced.

SETTING PRIORITIES FOR MANAGEMENT OF INVASIVE SPECIES

At this time, predicting the impacts of invasive species seems a lofty long-term goal, toward which we have made little progress. Although we should not be paralyzed by the apparent complexity of predicting invasions, perhaps a more realistic and optimistic short-term goal of invasion ecology lies in analyzing impacts of currently widespread invaders for broad taxonomic, trophic, or geographic patterns. Such a synthesis of empirical work may tell us something about how we should manage already widespread invasive species. For instance, do certain sorts of habitats suffer more damage from invaders than others, or do invasive predators tend to have stronger effects on recipient communities than invasive herbivores? The answers to these types of questions could eventually focus our efforts on the types of invaders or types of invaded communities that need the most immediate attention.

Meta-analysis, a technique gaining popularity in ecology (Gurevitch and Hedges, 1993), eventually may prove useful for synthesizing data gathered on impacts of invaders. This statistical technique takes advantage of multiple studies that address a similar question, in this case, the measured effect of the introduction of an invasive species into a system (Wonham et al., 2000). The strength of meta-analysis lies in its ability to combine many disparate studies to gain insight into very large-scale questions that would be beyond the scope of any individual research program. A meta-analysis on the impacts of invaders can address questions at a level beyond the interaction between an insect and its specific host plant, or between a species and a congeneric competitor. Therefore, it should be helpful in providing insight at the level of national policy regarding the risk of species introductions.

One of the frustrations of constructing a meta-analysis using the currently available data is that little concordance exists among studies in how results are communicated (Gurevitch et al., 1992), let alone how impact is measured (Wonham et al., 2000). At the very minimum, authors need to report basics, such as the number of replicates used in experiments, the unit of replication, and measures of variance around summary statistics. As we have alluded to above, the diversity of measures of impact make this task more difficult because the impact of an invader can depend on how it is quantified. A concerted effort among those researchers seriously studying the impact of an invasive species to employ a standard set of measures in their studies could benefit efforts to compare results from different studies and different systems.

For ecologists studying biological invasions, comparability among measures of impact allows the exploration of predictions about species interactions or the way that communities and ecosystems are structured. Comparability among measures has more than academic appeal, however. Those in the position to make decisions regarding regulation, control, or allocation of funding to solve problems generated by invasive species must often decide which problems to solve now and which can be passed over. It is hard to imagine how these decisions are made in the absence of information regarding the effects of the invaders in question. Ideally, we would like to rank invaders by their impact and choose the highest ranked cases as priority for management efforts. Recent work by Williamson (1998) explored correlations among measures of the impact of introduced weeds in Britain; he found strong concordance among some measures, but not others We may also obtain valuable predictions about the comparability among measures of impact from community models (Parker et al., 1999). Modeling, although a complement rather than a replacement for empirical work, has the advantage of being an efficient, low-cost technique for making unlimited numbers of comparisons and is not subject to the same logistical constraints as empirical studies.

CONCLUSIONS

In the United States, as throughout the world, our perception of sanitary and phytosanitary risk of species introductions and invasions has expanded from concern about agriculture and human health to concern about natural ecosystems and communities. As a result of this shift, regulatory bodies now must strive to incorporate appropriate levels of protection for natural systems, as well as agriculture and human health, in the development of new policy. Key to understanding the risk that invasive species pose to natural environments is understanding what types of ecological effects current invaders have on natural ecosystems. We have stressed that ecological impacts can take many forms and that sometimes they not only conflict with more anthropocentric impacts, but also with each other. Because some introductions and invasions by agricultural pests have been relatively well documented and well studied, they provide sound information about the biology of invaders, invaded systems, and the mechanisms of interaction between them from which to launch needed investigations into ecological impacts of invaders. Often these investigations of ecological impact follow earlier studies of economic or agricultural impact of the same invader, as was the case for the mosquito fish. We find it interesting that common problems plague efforts to quantify and predict the risk of invaders in both agricultural and ecological contexts: a poor understanding of the basic biology, ecology, and taxonomy of invasive, or potentially invasive species (especially for insects, pathogens, and marine invertebrates); inconsistency in data collection and reporting; and lack of complete, accessible databases of invaders. These problems existed in the nineteenth century when phylloxera jeopardized our chances of tasting a fine Bordeaux claret, and they still exist today as new pests, such as Asian long-horn beetles, monopolize media on environmental problems. Today, these deficiencies are particularly crippling because the rate of introductions has greatly accelerated due to increased human traffic (Lövel, 1997; Cohen and Carlton, 1998).

Several approaches may help both ecologists and policymakers make some sense of the complexity of ecological impacts. Focusing on vectors can help to define suites of invaders with similar biology and impacts and to increase our efficiency in evaluating their risk. In regulation, vectors may provide an efficient point of attack, as opposed to a species-by-species approach. Our ability to predict the establishment of certain types of invaders shows some promise and may be especially useful in the context of intentional introductions. As of yet, we are unable to predict which successful invasions may have the biggest impacts. Nevertheless, in the short term, it makes sense for managers, policymakers, and ecologists to work together in prioritizing control and prevention of current invaders by the impact they have on natural systems. Meanwhile, the difficult task of predicting impacts of potentially invasive organisms should continue through attempts to synthesize broad patterns of invader impacts (e.g., with meta-analysis), community modeling, and, of course, continued empirical research on as many systems as possible.

ACKNOWLEDGEMENTS

We are grateful for helpful comments and careful editing byJohn Hunter, Dan Simberloff, and anonymous reviewers. Some of the ideas presented here were developed in conjunction with the Invasions Working Group (1998–1999) that was supported by the National Center for Ecological Analysis and Synthesis, a center funded by the National Science Foundation (NSF) grant DEB-94-21535, the University of California at Santa Barbara, and the state of California. The completion of this paper also benefited from NSF grant DEB-98-01274 to K. Goodell, and NSF grant DEB-98-08501 to I. M. Parker, and NSF grant DEB-98-06517 to G. S. Gilbert. In addition, K. Goodell was supported by NSF grant DEB-97-07330 to J. D. Thomson during the preparation of this manuscript.

REFERENCES

Alderman, D.J. 1996. Geographical spread of bacterial and fungal diseases of crustaceans. Revue Scientifique et Technique International Office of Epizootics 15:603–632.

Anagnostakis, S.L. 1987. Chestnut Blight: the classical problem of an introduced pathogen. Mycologia 79(1):21–37.

AQIS (Australian Quarantine Inspection Service). 1999. PDI Database. Australian Department of Agriculture, Fisheries and Forestry, Canberra, Australia.

Arnaud, G., A. Rodriguez, A. Ortega-Rubio, and S. Alvarez-Cadenas. 1993. Predation by cats on the unique endemic lizard of Socorro Island (*Urosaurus auriculatus*), Revillagigedo, Mexico. Ohio Journal of Science 93:101–104.

Bergelson, J., and M. Crawley. 1989. Can we expect mathematical models to guide biological control programs? A comment based on case studies of weed control. Comments on Theoretical Biology 1:197–216.

Brown, L.R., and P.B. Moyle. 1991. Changes in habitat and microhabitat partitioning with an assemblage of stream fishes in response to predation by Sacramento squawfish (*Ptychocheilus grandis*). Canadian Journal of Fisheries and Aquatic Sciences 48:849–856.

Buchanan, G.A. 1987. The distribution of grape phylloxera, *Daktulosphaira vitifolii* (Fitch) in central and north-eastern Victoria. Australian Journal of Experimental Agriculture 27:591–595.

Carlton, J.T., and J.B. Gellar. 1993. Ecological roulette: the global transport of nonindigenous marine organisms. Science 261:78–82.

Carman, J.G., and J.D. Brotherson. 1982. Comparisons of sites infested and not infested with saltcedar (*Tamarix pentandra*) and Russian olive (*Eleagnus angustifolia*). Weed Science 30:360–364.

Ciccarone, A. 1949. Zonate leaf spot of sorghum in Venezuela. Phytopathology 39:760–761.

Cohen, A.N., and J.T. Carlton. 1998. Accelerating invasion rate in a highly invaded estuary. Science 279:555–558.

Cole, F.R., A.C. Medeiros, L.L. Loope, and W.W. Zuehlke. 1992. Effects of the Argentine ant on arthropod fauna of Hawaiian high-elevation shrubland. Ecology 73:1313–1322.

Constanza, R., R. d'Arge, R. de Groot, S. Farber, M. Grasso, B. Hannon, K. Limburg, S. Naeem, R. O'Neill, J. Paruelo, R.G. Raskins, P. Sutton, and M. van den Belt. 1997. The value of the world's ecosystem services and natural capital. Nature 387:253–260.

Cowling, E.B. 1978. Agricultural practices that favor epidemics. In J.G. Horsefall and E.B. Cowling, eds., Plant Disease: An Advanced Treatise, Vol. II. New York: Academic Press.

Craig, G.B.J. 1993. The diaspora of the Asian tiger mosquito. Pp. 101–120 in B. N. McKnight, ed., Biological Pollution: The Control and Impact of Invasive Exotic Species. Indianapolis: Indiana Academy of Sciences.

Daily, G.C., ed. 1997. Nature's Services. Washington, D.C.: Island Press.

D'Antonio, C.M., and B.E. Mahall. 1991. Root profiles and competition between the invasive, exotic perennial, *Carpobrotus edulia* and two native shrub species in California [USA] coastal scrub. American Journal of Botany 78:885–894.

Darwin C. 1859. On the Origins of Species by Means of Natural Selection, 1st ed. London: John Murray.

DeBach, P., and D. Rosen. 1991. Biological Control by Natural Enemies, 2nd ed. Cambridge, U.K.: Cambridge University Press.

Desdames, C. 1984. Victor Pulliat, viticulteur du beaujolais, vainqueur du phylloxera. Historia 454:77–78.

Diéguez-Uribeondo, J., C. Termiño, and J.L. Múzquiz. 1997. The crayfish plague fungus (*Aphanomyces astaci*) in Spain. Bulletin Français de la Pêche et de la Piciculture 347:753–763.

Dombeck, M. 1996. Noxious and invasive weeds: should we have a national policy for a national problem? Proceedings of the Western Society of Weed Science 49:5–9.

Drake, J.A. 1983. Invasibility in Lotka-Volterra interaction webs. In D. D'Angelis, W.M. Post, and G. Siguhara, eds., Current Trends in Food Web Theory. Oak Ridge, Tenn.: Oak Ridge National Laboratory.

Ebert, D. 1998. Evolution-experimental evolution of parasites. Science 282:1432–1435.

Echelle, A.A., and P.J. Conner. 1989. Rapid, geographically extensive genetic introgression after secondary contact between two pupfish species (Cyprinodon, Cyprinodontidae). Evolution 43:717–727.

Ellstrand, N.C., and C.A. Hoffman. 1990. Hybridization as an avenue of escape for engineered genes—strategies for risk reduction. Bioscience 40:438–442.

Elton, C.S. 1958. Ecology of Invasions by Plants and Animals., London: Metheun.

Ewald, P.W. 1994. Evolution of Infectious Diseases. Oxford, U.K.: Oxford University Press.

Forcella, F., J.T. Wood, and S.P. Dillon. 1986. Characteristics distinguishing invasive weeds within *Echium* (Bugloss). Weed Research 26:351–364.

Fritts, T.H. 1999. A summary of documented arrivals of Brown Tree snakes (and sightings on Saipan likely to represent Brown Tree snakes) to islands and the continental United States. U.S. Geological Survey, Patuxent Wildlife Research Center. Available online at http//www.pwrc.nbs.gov/btdisp2.htm.

Fritts, T.H., and G.H. Rodda. 1998. The role of introduced species in the degradation of island ecosystems: a case history of Guam. Annual Review of Ecology and Systematics 29:113–140.

Gause, G.F. 1934. The Struggle for Existence., Baltimore, Md: Williams & Wilkins (reprinted in 1964 by Hafner, New York).

Gentle, C.B., and J.A. Duggin. 1997. Allelopathy as a competitive strategy in persistent thickets of *Lantana camara* L. in three Australian forest communities. Plant Ecology 132:85–95.

Gerberich, J.B. 1946. An annotated bibliography of papers relating to the control of mosquitoes by the use of fish. The American Midland Naturalist 36:87–131.

Goodell, K. 1998. Impacts of introduced honeybees on native solitary bees. Bulletin of the Ecological Society of America Abstracts: 62.

Goodell, K., M. Wonham, B. Von Holle, and I.M. Parker. 2000. Trophic and taxonomic patterns of impact of invasive species: a meta-analysis. Bulletin of the Ecological Society of America Abstracts. In press.

Granett, J., P. Timper, and L.A. Lider. 1985. Grape phylloxera (*Daktulosphaira vitifoliae*) (Homoptera:Phylloxeridae) biotypes in California. Journal of Economic Entomology 78:1463–1467.

Guan, R.-Z., and P.R. Wiles. 1996. Ecological impact of introduced crayfish on benthic fishes in a British lowland river. Conservation Biology 11:641–647.

Gurevitch, J., and L.V. Hedges. 1993. Meta-analysis: combining the results of independent experiments. Pp. 378–398 in S. M. Scheiner and J. Gurevitch, eds., Design and Analysis of Ecological Experiments. New York: Chapman & Hall.

Gurevitch, J., L.L. Morrow, A. Wallace, and J.S. Walsh. 1992. A meta-analysis of competition in field experiments. American Naturalist 140:539–572.

Haack, R.A., K.R. Law, V.C. Mastro, H.S.Ossenbruggen, and B.J. Raimo. 1997. New York's battle with the Asian long-horned beetle. Journal of Forestry 95(12):11–15.

Hager, H.A., and K.D. McCoy. 1998. The implications of accepting untested hypotheses: a review of the effects of purple loostrife (*Lythrum salicaria*) in North America. Biodiversity and Conservation 7:1069–1079.

Hedges, L.V. 1996. Statistical considerations. Pp. 29–38 In H. Cooper and L. V. Hedges, eds., The Handbook of Research Synthesis. New York: Russell Sage Foundation.

Hedrick, R.P., M. El-Matbouli, M.A. Adkison, and E. MacConnell. 1998. Whirling Disease: reemergence among wild trout. Immunological Reviews 166:365–376.

Hoffman, G.L. 1970. Intercontinental and transcontinental dissemination and transfaunation of fish parasites with emphasis on whirling disease (Myxosoma cerebralis). Pp. 69–81 in Diseases of Fish and Shellfish, S.F. Snieszko, ed. Washington, D.C.: American Fisheries Society.

Holland, E.A. 1933. An experimental control of malaria in New Ireland by distribution of Gambusia affinis. Transactions of the Royal Society of Tropical Medicine and Hygiene 26:529–538.

Howard, H.H. 1922. An indigenous fish used in combating malaria. Nation's Health 4:65–69, 139–143.

Howe, E., C. Howe, R. Lim, and M. Burchett. 1997. Impact of the introduced poeciliid Gambusia holbrooki (Girard, 1859) on the growth and reproduction of Pseudomugil signifer (Kner, 1865) in Australia. Marine and Freshwater Research 48:425–434.

Hurlburt, S.H., J. Zelder, and D. Fairbanks. 1972. Ecosystem alteration by mosquitofish (Gambusia affinis) predation. Science 175:639–641.

Jarosz, A.M., and A.L. Davelos. 1995. Tansley Review No. 81: Effects of disease in wild plant populations and the evolution of pathogen aggressiveness. The New Phytologist 129:834–841.

Kato, M.; A. Shibata, T. Yasui, and H. Nagamasu. 1999. Impact of introduced honeybees, Apis mellifera upon native bee communities in the Bonin (Ogasawara) Islands. Res. Popul. Ecol. 41:217–228.

King, P.D., and G.A. Buchanan. 1986. The dispersal of phylloxera crawlers and spread of phylloxera infestations in New Zealand and Australian vineyards. American Journal of Enology and Viticulture 37:26–33.

Krumholz, L.A. 1948. Reproduction in the western mosquito fish Gambusia affinis affinis (Baird and Girand) and its use in mosquito control. Ecological Monographs 18:1–43.

Langham, N.P.E. 1990. The diet of feral cats (Felis catus) on Hawke's Bay farmland, New Zealand. New Zealand Journal of Zoology 17:243–256.

Laurent, P.J. 1997. Introductions d'ecrevisses en France et dans le monde, historique et conséquences. Bulletin Français de la Pêche et de la Piciculture 344/345:345–356.

Lilley, J.H., L. Cerenius, and K. Söderhäll. 1997. RAPD evidence for the origin of crayfish plague outbreaks in Britain. Aquaculture 157:181–185.

Lloyd, L. 1989. The ecological implications of Gambusia holbrooki with Australian native fishes. Pp. 94–97 in D. A. Pollard, ed., Introduced and Translocated Fishes and their Ecological Effects. Canberra: Australian Government Publishing Service.

Lloyd, L., A.H. Arthington, and D.A. Milton. 1986. The mosquito fish—a valuable mosquito-control agent or a pest? Pp. 5–25 in R. L. Kitching, ed., The Ecology of Exotic Animals and Plants: Some Australian Case Histories. New York: John Wiley & Sons.

Locke, A., D.M. Reid, H.C. VanLeeuwen, W.G. Sprules, J.T. Carlton. 1993. Ballast water exchange as a means of controlling dispersal of fresh-water organisms by ships. Canadian Journal of Fisheries and Aquatic Sciences 50(10):2086–2093.

Louda, S.M. 1998. Population growth of Rhinocyllus cinicus (Coleoptera:Curculionidae) on two species of native thistles in prairie. Environmental Entomology 27:834–841.

Louda, S.M., D. Kendall, J. Connor, and D. Simberloff. 1997. Ecological effects of an insect introduced for the biological control of weeds. Science 277:1088–1090.

Lövel, G. 1997. Global change through invasion. Nature 388:627–628.

Lowery, R.S., and D.M. Holdich. 1988. *Pacifasticus leniusculus* in North America and Europe, with details of the distribution of introduced and native crayfish species in Europe. Pp. 283–308 in D.M. Holdich and R.S. Lowery, eds., Freshwater Crayfish: Biology, Management and Exploitation. London: Croom Helm Ltd.

Lukacs, P. 1996. How America saved Europe's vineyards. American Heritage 47:89.

Macisaac, H.J. 1996. Potential abiotic and biotic impacts of zebra mussels on the island waters of North America. American Zoologist 36:287–299.

Mack, M.C., and C.M. D'Antonio. 1998. Impacts of biological invasions on disturbance regimes. Trends in Ecology and Evolution 13:195–198.

Mack, R.N. 1996. Predicting the identity and fate of plant invaders: emergent and emerging approaches. Biological Conservation 78:107–121.

Matthews, M., and J.D. Reynolds. 1992. Ecological impact of crayfish plague in Ireland. Hydrobiologia 234:1–6.

McEvoy, P.B., and E.M. Coombs. 1999. A parsimonious approach to biological control of plant invaders. Ecological Applications 9:387–401.

McKay, R.J. 1984. Introductions of exotic fishes in Australia. Pp. 177–199 in W.R. Courtenay, Jr. and J. Stauffer Jr., eds., Distribution, Biology, and Management of Exotic Fishes. Baltimore, Md: Johns Hopkins University Press.

McMillan, M., and D. Wilcove. 1994. Gone but not forgotten: why have species protected by the Endangered Species Act become extinct? Endangered Species Update 11:5–6.

Mills, E.L., J.H. Leach, J.T. Carlton, and C.L. Secor. 1993. Exotic species in the Great Lakes: a history of biotic crisis and anthropogenic introductions. Journal of Great Lakes Research 19:1–54.

Morgan, L.A., and W.A. Buttemer. 1996. Predation by the non-native fish *Gambusia holbrooki* on small *Littoria aurea* and *L. dentata* tadpoles. Australian Zoologist 30:143–149.

Morton, L.T. 1985. Winegrowing in Eastern America., Ithaca, N.Y.: Cornell University Press.

Neill, W. 1983. The tamarix invasion of desert riparian areas. Educational Bulletin of the Desert Protective Council, Vol. 83–84. Spring Valley, Calif: Educational Foundation of the Desert Protective Council.

Nylund, V., and K. Westman. 1992. Crayfish diseases and their control in Finland. Finnish Fisheries Research 14:107–118.

Oestreicher, A. 1996. La crisis filoxérica en España (estudio comparativo sobre las consecuencias socio-económicas de la filoxera en algunas regiones vitivinícolas Españolas). Hispania 1/2:587–622.

O'Neill, C.R. 1996. Economic Impact of Zebra Mussels: the 1995 National Zebra Mussel Information Clearinghouse Study., Ithaca, N.Y.: Cornell Cooperative Extension.

Palm, M.E. 1999. Mycology and world trade: a review from the front line. Mycologia 91:1–12.

Parker, I.M., D. Simberloff, W.M. Lonsdale, K. Goodell, M. Wonham, P.M. Kareiva, M.H. Williamson, B. Von Holle, P.B. Moyle, J.E. Byers, and L. Goldwasser. 1999. Impact: toward a framework for understanding the ecological effects of invaders. Biological Invasions 1(1):3–19.

Paton, D.C. 1997. Honeybees *Apis mellifera* and the disruption of plant-pollinator systems in Australia. Victorian Naturalist 114:23–29.

Pearre, S., and R. Maass. 1998. Trends in the prey size-based trophic niches of feral and house cats *Felis catus* L. Mammal Review 28:125–139.

Perrins, J., M. Williamson, and A. Fitter. 1992. A survey of differing views of weed classification: implications for regulation of introductions. Biological Conservation 60:47–56.

Porter, S.D., and D.A. Savignano. 1990. Invasion of polygyne fire ants decimates native ants and disrupts arthropod community. Ecology 71:2095–2106.

Pouget, R. 1990. Histoire de la Lutte Contre le Phylloxera de la Vigne en France. Versailles: Institut Nacional de la Recherche Agronomique.

Power, M.E., D. Tilman, J.A. Estes, B.A. Menge, W.J. Bond, L.S. Mills, G. Daily, J.C. Castilla, J. Lubchenco, and R.T. Paine. 1996. Challenges in the quest for keystones: identifying keystone species is difficult—but essential to understanding how loss of species will affect ecosystems. BioScience 46:609–620.

Reichard, S.H., and C.W. Hamilton. 1997. Predicting invasions of woody plants introduced into North America. Conservation Biology 11:193–203.

Rejmánek, M., and D.M. Richardson. 1996. What attributes make some plant species more invasive? Ecology 77:1655–1661.

Rhymer, J.M., and D. Simberloff. 1996. Extinction by hybridization and introgression. Annual Review of Ecology and Systematics 27:83–109.

Ricciardi, A., and H.J. MacIsaac. 2000. Recent mass invasion of the North American Great Lakes by Ponto-Caspian species. Trends in Ecology and Evolution 15(2):62–65.

Richardson, D.M., R.M. Cowling, and D.C. Le Maitre. 1990. Assessing the risk of invasive species in *Pinus* and *Banksia* in South African mountain fynbos. Journal of Vegetation Science 1:629–642.

Robinson, W.S, R. Nowogrodzki, and R.A Morse. 1989. The value of honeybees as pollinators of U.S. crops. American Bee Journal 129:411–423.

Roubik, D.W. 1982. Ecological impact of Africanized honeybees on native neotropical pollinators. Pp. 233–247 in: Social Insects in the Tropics, Vol. 1., P. Jaisson, ed. Paris, France: Université de Paris.

Ruesink, J.L., I.M. Parker, M.J. Groom, and P.M. Kareiva. 1995. Reducing the risks of nonindigenous species introductions: guilty until proven innocent. BioScience 45:465–477.

Savidge, J.A. 1987. Extinction of an island avifauna by an introduced snake. Ecology 68:660–668.

Schmitz, D.C., D. Simberloff, R.H. Hofstetter, W. Haller, and D. Sutton. 1997. The ecological impact of nonindigenous plants. Pp. 39–61 in D. Simberloff, D.C. Schmitz, and T.C. Brown, eds., Strangers in Paradise: Impact and Management of Nonindigenous Species in Florida., Washington, D.C.: Island Press.

Scott, D.J. 1971. The importance to New Zealand of seedborne infection of *Helminthosporium maydis*. Plant Disease Reporter 55:966–968.

Shearer, B.L., and M. Dillon. 1995. Susceptibility of plant species in *Eucalyptus marginata* forests to infection by *Phytphthora cinnimomi*. Australian Journal of Botany 43:113–134.

Simberloff, D. 1981. Community effects of introduced species. Pp. 53–81 in M.H. Nitecki, ed., Biotic Crises in Ecological and Evolutionary Time. New York: Academic Press.

Simberloff, D. 1986. Introduced insects: a biogeographic and systematic perspective. Pp. 3–26, in: Mooney, H.A. and J.A. Drake, eds. Ecology of Biological Invasions of North America and Hawaii. Springer-Verlag, Berlin.

Simberloff, D. 1991. Keystone species and community effects of biological introductions. In L.R. Ginzburg, ed., Assessing Ecological Risks of Biotechnology. Boston, Mass.: Butterworth-Heinemann.

Simberloff, D., and P. Stiling. 1996a. How risky is biological control? Ecology 77:1965–1974.

Simberloff, D., and P. Stiling. 1996b. Risks of species introduced for biological control. Biological Conservation 78:185–192.

Simberloff, D., and B. Von Holle. 1999. Positive interactions of nonindigenous species: invasional meltdown? Biological Invasions 1(1):21–32.

Söderbäck, B. 1994. Interactions among juveniles of two freshwater crayfish species and a predatory fish. Oecologia 100:229–235.

Stevenson, I. 1980. The diffusion of disaster: the phylloxera outbreak in the *département* of the Hérault, 1862–1880. Journal of Historical Geography 6:47–63.

Strong, D.R., J.H. Lawton, and T.R.E. Southwood. 1984. Insects on Plants. Harvard. Cambridge, Mass: University Press.

Taugbol, T., and J. Skurdal. 1993. Crayfish plague and management strategies in Norway. Biological Conservation 63:75–82.

Thomas, W. 1973. Seed-transmitted squash mosaic virus. New Zealand Journal of Agricultural Research 16:561–567.

Thompson, J.N. 1998. Rapid evolution as an ecological process. Trends in Ecology and Evolution 13:329–332.

Unestam, T. 1973. Significance of diseases on freshwater crayfish. International Symposium on Freshwater Crayfish 1:135–150.

U.S. Coast Guard. 1993. Ballast water management for vessels entering the Great Lakes. Code for Federal Regulations 33-CFR Part 151.1510. Effective May 10, 1993.

U.S. Congress. 1993. Harmful Nonindigenous Species in the United States. Washington D.C.: Office of Technology Assessment.

USDA (U.S. Department of Agriculture). 1991. Pest Risk Assessment of the Importation of Larch from Siberia and the Soviet Far East, Vol. Publ. No. 1495. Washington, D.C.: U.S. Government Printing Office.

USDA (U.S. Department of Agriculture), Agricultural Research Service. 1999. National Genetic Resources Program. Germplasm Resources Information Network (GRIN). Online Database National Germplasm. http://www.ars-grin.gov.

Van der Zwet, T., and S.V. Beer. 1992. Fire Blight—Its Nature, Prevention, and Control. A Practical Guide to Integrated Disease Management. USDA Agricultural Information Bulletin No. 631. Washington, D.C.: Government Printing Office.

Van Zyl, D.J. 1984. *Phylloxera vastatrix* in die kaapkolonie, 1886–1900: voorkoms, verspreiding en ekonomiese. South African Historical Journal 16:26–48.

Vitousek, P.M., and L.R. Walker. 1989. Biological invasions by *Myrica faya* in Hawaii: Plant demography, nitrogen fixation, ecosystem effects. Ecological Monographs 59:247–265.

Vitousek, P.M., L.R. Walker, L.D. Whiteaker, D. Mueller-Dombios, and P.A. Matson. 1987. Biological invasion by *Myrica faya* alters ecosystem development in Hawaii. Science 238:802–804.

von Broembsen, S.L. 1989. Invasions of natural ecosystems by plant pathogens. Pp. 77–83 in J.A. Drake, H.A. Mooney, F. di Castri, R.H. Groves, F.J. Kruger, M. Rejmánek, and M. Williamson, eds., Biological Invasions: A Global Perspective. Scientific Committee on Problems of the Environment (SCOPE) of the International Council of Scientific Unions. New York: John Wiley & Sons.

Waage, J.K., and D.J. Greathead. 1988. Biological control and opportunities. Philosophical Transactions of the Royal Society of London B. 318:111–128.

Webb, C., and J. Joss. 1997. Does predation by the fish *Gambusia holbrooki* (Atheriniformes: Poeciliidae) contribute to declining frog populations? Australian Zoologist 30:316–324.

Werner, I., and J.T. Hollibaugh. 1993. Comparison of clearance rates and assimilation efficiency for phytoplankton and bacterioplankton. Limnology and Oceanography 38:949–964.

Williamson, M. 1996. Biological Invasions. London: Chapman & Hall.

Williamson, M. 1998. Measuring the impact of plant invaders in Britain. Pp. 57–68 in U. Starfinger, K. Edwards, I. Kowarik, and M. Williamson, eds., Plant Invasions: Ecological Consequences and Human Responses. Leiden, The Netherlands: Backhuys.

Wonham, M., K. Goodell, B. Von Holle, and I.M. Parker. 2000. Ecological and geographic patterns of impact in invaded communities: a meta-analysis. Bulletin of the Ecological Society of America Abstracts. In press.

6

Risk Management and the World Trading System: Regulating International Trade Distortions Caused by National Sanitary and Phytosanitary Policies

DAVID G. VICTOR

Council on Foreign Relations, New York

The Sanitary and Phytosanitary (SPS) Agreement, part of the 1994 accords that established the World Trade Organization (WTO), promotes international trade by requiring countries to base their sanitary (human and animal safety) and phytosanitary (plant safety) measures on international standards. However, it allows countries wide latitude to deviate from international standards when choosing their level of SPS protection, provided that (1) countries base their deviations on scientific risk assessment, (2) countries avoid discrimination by requiring comparable levels of SPS protection in comparable situations, and (3) countries not implement SPS measures that are more restrictive of trade than necessary to achieve the level of SPS protection that they seek. In this paper I review and assess the major provisions of the SPS Agreement (Appendix A), the international SPS standard-setting bodies, and the disciplines that govern allowable deviations from those international standards. I also examine the three WTO disputes that have helped to interpret the provisions of the SPS Agreement: the European Community's (EC)[1] ban on meat produced using growth hormones, Australia's ban on imports of fresh and frozen salmon from Canada, and Japan's fumigation testing requirements for imported fruits and nuts.

---

[1]For simplicity, I refer to the European Community, which today is also often called the European Union, as the EC.

Although disputes have not led to full interpretation of the major provisions of the agreement, it appears that the SPS Agreement has not led to the "harmonizing down" of SPS protection that many opponents of free trade have feared. Instead, the wide latitude permitted by the SPS Agreement has allowed national diversity in SPS measures to thrive while also reducing barriers to trade. International standards have not become a straitjacket—rather, they have had remarkably little impact on national SPS protection policies. (The main exceptions are in countries, especially in the developing world, that have not already adopted elaborate SPS protection policies; for those countries, international standards fill gaps and raise—not lower—the level of SPS protection.) The main impact of the agreement appears to be in harmonizing the process by which nations set SPS policies—notably, it is promoting greater use of risk assessment at the national level. More extensive assessment of risks may actually yield greater diversity in national SPS policies. In the paper I also suggest that the novel mechanisms for providing expert advice to WTO dispute panels have been highly effective and have greatly reduced the problems of "advocacy science" that often plague the use of risk assessment in other judicial proceedings. The story—apparent success in imposing international discipline that promotes trade while accommodating national diversity—may be a useful guide for solving similar problems that are the mainstay of the "trade and environment" debate.

INTRODUCTION

One measure of the success of the postwar trading system is that tariff trade barriers have declined sharply. But the reduction in tariffs has exposed the many nontariff barriers that remain, and in many cases governments have kept protectionism in place by simply shifting from tariff to nontariff measures. Included in the broad category of nontariff barriers are differences in technical standards such as labeling requirements and environmental regulations. The focus in this paper is on one subset of these technical barriers: measures for sanitary (animal, including human) and phytosanitary (plant) protection.

SPS measures often have huge effects on trade; yet managing them is not easy. SPS measures vary across and within nations because preferences and circumstances vary. Some nations seek tight protection while others readily consume riskier foods; some pristine environments are vulnerable to pest infestations and require elaborate quarantines for imported products, but other countries are already overrun with pests. The political and technical challenge for advocates of free trade is to accommodate such differences while stripping away SPS measures that are merely disguised protectionism.

In this paper I examine the effectiveness of the 1994 WTO Agreement on the Application of Sanitary and Phytosanitary Measures (SPS Agreement), which is the most significant global effort to reduce trade distortions caused by differences in national SPS protection policies. I examine the major elements of the SPS Agreement and the three international SPS standard-setting processes that are explicitly mentioned in the SPS Agreement. I briefly consider two other

WTO agreements—the General Agreement on Tariffs and Trade 1994 (GATT) and the Agreement on Technical Barriers to Trade (TBT) that are often invoked, along with the SPS Agreement, in studies that examine how the international trading system attempts to accommodate differences in national regulations. I review the major elements and decisions of the three WTO disputes that have concerned SPS measures, which help reveal how the WTO system is interpreting the SPS Agreement. And I identify major conclusions that can be drawn about the operation of this system. Throughout, the goal is not only to assess the SPS Agreement but also to explore the policy question that arises wherever expanding the scope of free trade rules intrudes into national policy: Can international rules and institutions impose discipline on national policy without requiring harmonization to international standards? That question arises frequently—especially in the debate over "trade and environment"—and the SPS Agreement demonstrates a slightly positive answer.

THE SPS AGREEMENT: MAJOR ELEMENTS

The basic obligations for members of the world trading regime have not changed since the first GATT agreement in 1947: Members must give equal treatment to exports from all members, and members are barred from discriminating between locally produced and imported products. Exceptions were allowed for tariffs on specific products, that were "bound" at specific levels. Numerous other "general exceptions" were also allowed for many national policy purposes, such as protection of human, animal, or plant life or the conservation of exhaustible natural resources. But those general exceptions—listed in the famous Article XX—were described only briefly. A system of "dispute panels" emerged to handle conflicts. In principle, the dispute panel system could have clarified the scope of Article XX. But in practice any GATT member could block adoption of a GATT panel report; and the panel system was often inactive, erratic in operation, and ineffective in major cases.[2] Enforcement that did exist was mainly through reciprocity imposed by GATT members themselves. But the blunt instrument of unilateral reciprocity was poorly suited for working out and applying the complex legal interpretations that would be needed to make Article XX workable. In the early decades of the GATT, tariffs were the largest barriers to trade. The main result from each of the first six rounds of negotiations to strengthen the GATT was to revise the list of tariff bindings and reduce the tariff impact on trade. Nontariff measures remained in shadow.

For the past 30 years, attention to nontariff measures has grown. The 1979 Tokyo Round agreements, which resulted from the seventh round of negotiations, included a separate "standards code" that imposed discipline on technical barriers to trade. But the code, like the GATT agreement, was backed by little enforcement; although all GATT members were bound by the GATT's

---

[2]For a comprehensive treatment of the cases that were handled, see Hudec (1993).

core rules, they were largely free to pick and choose among "code" rules. The result of the Tokyo Round's "GATT a la carte," most experts agree, had little effect on lowering technical barriers to trade.

The failures of earlier efforts were addressed head-on in the most recent (eighth) Uruguay Round of negotiations. By 1986, the year that the Uruguay Round began, nearly 90 percent of U.S. food imports were affected by nontariff barriers to trade, up from only half in 1966 (Tutwiler, 1991, cited in Vogel, 1995).[3] Exporters had a growing interest in taming these barriers.

The main legal products of the Uruguay Round were adopted in 1994. They were an updated version of the GATT (1994) along with 14 other agreements on textiles, subsidies, technical barriers to trade, SPS measures, and other topics. The Uruguay Round also produced a stronger and more judicial dispute-resolution procedure in which three-person panels hear and decide disputes and a standing Appellate Body hears appeals, and produced a mechanism that reviews trade policy in all member countries on a regular basis. Together, these agreements form a single, integrated package of obligations that constitutes the core obligations of a new international organization: The World Trade Organization.[4] Countries were no longer free to pick and choose their free trade commitments.

The most important element of the WTO concerning SPS protection is the Agreement on the Application of Sanitary and Phytosanitary Measures (SPS Agreement). The agreement's central purpose is to promote international trade by limiting the use of SPS measures as disguised barriers to trade. The agreement's basic rights and obligations (Article 2) underscore that WTO members have the right to impose SPS measures as necessary "for the protection of human, animal or plant life or health" (Articles 2.1 and 2.2). But members may not arbitrarily or unjustifiably discriminate between members; nor may members use SPS measures as disguised barriers to trade (Article 2.3). These basic rights and obligations are quite general, and thus efforts to interpret them have focused on the more detailed provisions of the SPS Agreement (in particular Article 5, which is detailed below).

In addition to restraining the SPS policies that countries may develop on their own, the SPS Agreement urges members to implement international standards. The agreement's preamble underscores the goal: *"Desiring to further*

---

[3] For a current overview of all technical barriers to trade in U.S. agriculture exports see Roberts and DeRemer, (1997).

[4]In addition, the WTO agreement included four "plurilateral" agreements (on aircraft, government procurement, dairy products, and bovine meat) that were adopted in 1994 along with the core WTO agreements. Unlike the "multilateral" obligations that all WTO members must implement, plurilateral agreements are optional. They are not necessarily useless because an agreement—even if voluntary—helps to signal proper conduct and facilitate cooperation. Moreover, voluntary agreements often lay the groundwork for later agreements that are binding and backed by an enforcement mechanism. For example, the conclusion of the seventh round in 1979 included a plurilateral code on technical barriers to trade; the failure of that code to have much effect led to the creation of similar, but binding, multilateral TBT and SPS agreements that were adopted in 1994 along with the other WTO agreements.

the use of harmonized sanitary and phytosanitary measures between Members, on the basis of international standards, guidelines and recommendations developed by the relevant international organizations...." The agreement declares that "Members shall base their sanitary and phytosanitary measures on international standards, guidelines or recommendations...." (Article 3.1). When a member imposes SPS measures that conform with international standards, guidelines, or recommendations, those measures will automatically be "presumed to be consistent with the relevant provisions of this Agreement...." (Article 3.2). However, countries may introduce measures that are stricter than international standards "if there is a scientific justification, or as a consequence of the level of [SPS] protection a Member determines to be appropriate in accordance with the relevant provisions ...of Article 5."[5]

Thus WTO members face a choice. A member may simply implement international standards,[6] where they exist, or deviate from those standards. To examine how the agreement affects the SPS measures that countries implement, it is thus necessary to examine both outcomes: (1) how international standards are established, and (2) the exceptions that permit a country[7] to deviate from

---

[5]The SPS agreement also includes a footnote at this point: "For the purposes of paragraph 3 of Article 3, there is a scientific justification if, on the basis of an examination and evaluation of available scientific information in conformity with the relevant provisions of this Agreement, a Member determines that the relevant international standards, guidelines or recommendations are not sufficient to achieve its appropriate level of sanitary or phytosanitary protection." Although the obligations and reasoning are a bit convoluted, this footnote has been interpreted as meaning that measures that deviate from international standards are acceptable if based on a risk assessment—that is, if they meet the requirements of Article 5, which includes the requirement of a risk assessment (Article 5.1). In plain language, Article 3 promotes harmonization with international standards. And Article 5 allows countries to escape the straitjacket of international standards, provided that an assessment of risks is the first step in setting such stricter SPS measures.

[6]For simplicity, hereafter I use the term "international standards" to denote "international standards, guidelines, or recommendations." Although the full term is important for legal purposes because it is broader, the simpler plain English term is most appropriate for this paper. One of the remaining gray zones in applying the agreement concerns just how broadly to apply this definition. For example, as I review below, the *Codex Alimentarius* Commission adopts not only specific standards (e.g., on food additives) but also more general standards for commodities and advisory guidelines. Does the WTO Agreement apply to all three, even though *Codex* guidelines were never designed nor intended to have binding application?

[7]For simplicity I use the terms "country" and "WTO member" interchangeably. For purposes of discussing legal obligations I also treat countries as single units. However, some SPS measures (e.g., quarantines) apply only to certain parts of countries and thus have trade effects only for imports (from outside as well as inside the country) into that part of the country. Examples include quarantines for many exports to Hawaii, which are stricter than exports to the rest of the United States. Moreover, although the obligations of the WTO agreements are imposed on "members," it is not necessary that *governments* perform all of the required tasks. Often risk assessments and trade controls are

those international standards. I address these in reverse order because the exceptions are the most elaborate portion of the SPS Agreement and all of the disputes involving the SPS Agreement have focused on how to interpret the exceptions. If a country implements an international standard, it is automatically in compliance with the SPS Agreement, and thus all the WTO disputes concern instances where either international standards are absent, or a member has chosen not to implement existing standards.

Before turning to international standards and exceptions, it is important to note that the SPS Agreement includes several important obligations that extend the agreement's influence beyond simply the setting of SPS levels and measures. In principle, the SPS Agreement also allows exporters broad latitude when determining the SPS measures that are needed to meet the level of SPS protection that importers demand. The agreement requires that importers accept the SPS measures of exporters...

> ...as equivalent, even if these measures differ from their own or from those used by other Members trading in the same product, if the exporting Member objectively demonstrates to the importing Member that its measures achieve the importing Member's appropriate level of [SPS] protection (Article 4.1).[8]

Assuming that exporters have an interest in identifying the least trade-restrictive measure, this "equivalence" requirement could automatically ensure that SPS rules are not more discriminatory than necessary; "equivalence" could also open markets without requiring actual harmonization. In another context—the creation of the EC's single market—similar concepts (e.g., "mutual recognition") created a strong market-opening dynamic by allowing legal production from any European country into any other European national market. The agreement also requires that countries make their SPS policies transparent both through publication and creation of national "enquiry points" that can answer any reasonable question about that country's SPS rules (Articles 5.8 and 7, and Annex B). If that system operates properly then exporters will find it easier to comply with an importer's SPS rules, which should promote trade. Transparency is also essential to making use of the equivalence requirement described above. In addition, the agreement creates an international SPS Committee that meets on a regular basis to consider relevant topics and periodically review the performance of the SPS Agreement (Article 12). That committee is expected to adopt guidelines on SPS-related issues that could help in the interpretation of the agreement, although, to date, its impact on trade patterns has been minimal.

---

implemented by nongovernmental organizations (especially private firms, industrial associations and scientific laboratories), with government acting only a supervisor (see SPS Agreement, Article 13).

[8]The SPS Agreement also includes a specific application of the "equivalent" requirement, which is especially important for SPS measures: pest- and disease-free areas. Countries that can demonstrate that all or some of their country is free from a hazard are allowed to circumvent SPS measures that are intended to block diseases on products from that country (Article 6).

The agreement allows the least developed countries to delay implementation of the agreement for five years (Article 14), allows other extensions, and empowers the SPS Committee to grant temporary extensions and relief from the agreement's obligations in cases of hardship.

The Exceptions

One of the most controversial aspects of the debate over opening trade has been the fear that free trade will force all countries to harmonize their national standards into a straitjacket of international standards. Donning the straitjacket, skeptics argue, could force nations to adopt stricter SPS measures than they would otherwise want. That might force societies to spend resources on SPS protection that they could have devoted to other purposes such as economic development. Or the straitjacket could force countries that already have tight SPS measures to relax them, leading perhaps to downward harmonization if international standards merely mirror the lowest common denominator. The latter has been the most controversial because existing SPS measures are generally much tighter in the advanced industrialized countries, which is also where most of the public interest groups active on SPS issues are located. Harmonization, they fear, will require compromising hard-won rules that protect consumers and the environment (Silverglade, 1998; Jacobsen, 1997).[9]

Because of this heated debate, fully under way when the WTO agreements were negotiated, the SPS Agreement permits countries to adopt SPS protection policies that are stricter or weaker than international standards. Rather than requiring harmonization, the SPS Agreement imposes discipline on both the *level* of SPS protection that countries seek and the *measures* they impose to attain those levels. The agreement and disputes over interpretation of the agreement have underscored that any country may set the *level* of SPS protection that it determines to be "appropriate." (This "appropriate level" is also often termed in the literature on risk management as the "level of acceptable risk.") The SPS Agreement does impose some discipline on the level of SPS protection, but it imposes more elaborate discipline on the measures that countries use to achieve that level. Below, I address the disciplines imposed on SPS levels and measures that are stricter than the international standards, and then I discuss measures that are weaker.

SPS Levels and Measures that are Stricter than the International Standard

The SPS Agreement is mainly intended to discipline SPS measures that cause an unjustified barrier or restriction on trade because they are stricter than international standards. Indeed, Article 3.3 (cited above) explicitly carves out an

---

[9]Also, there have been numerous letters to the President of the United States, responses to proposed rule making, and other political actions based on similar arguments.

exception to the goal of harmonization for SPS measures that are stricter than international standards. Article 3.3 requires that a member must be able to provide "scientific justification" for choosing a higher level of SPS protection. Similarly, Article 2.2 requires that members base their SPS measures on "scientific principles." These general requirements are quite broad and thus, in practice, the decisions of the Panels and Appellate Body in the three WTO disputes related to the SPS Agreement have turned to Article 5 for a more detailed description of "scientific" determination of SPS levels and measures.[10]

Article 5 requires that SPS measures be "based on an assessment, as appropriate to the circumstances, of the risks to human, animal or plant life or health, *taking into account risk assessment techniques developed by the relevant international organizations*" (Article 5.1, emphasis added). It requires that members take into account available scientific evidence (Article 5.2). When performing risk assessments, countries must account for economic factors such as potential loss in production or sales if a pest or disease enters the country as well as the cost effectiveness of different measures that could limit such risks (Article 5.3).

Article 5 also underscores that the agreement does not address every aspect of SPS protection. Rather, it concerns principally those SPS policies that affect trade. It urges countries to minimize the negative trade effects of SPS measures (Article 5.4). It requires that countries avoid "arbitrary or unjustifiable distinctions" in their levels of SPS protection "*if such distinctions result in discrimination or a disguised restriction on international trade*" (Article 5.5, emphasis added). Article 5.6 requires that countries not impose SPS measures that are "more trade-restrictive than required to achieve [the level of SPS protection that the member deems appropriate]." A footnote to Article 5.6 declares that a measure would be inconsistent with Article 5.6 if an alternative is found that passes each of the following three tests: (a) it is "reasonably available," (b) it achieves the member's appropriate level of SPS protection, and (c) it is "significantly less restrictive to trade than the SPS measure contested." Article 5.7 allows countries to adopt SPS measures even in the absence of good scientific information, provided that they also establish a process to obtain the information needed for a proper risk assessment.

---

[10]The legal reasoning is a bit convoluted because the SPS Agreement is also convoluted on this point. Article 3.3 specifically cites Article 5 as a justification for countries to deviate from international standards. (However, the citation is odd because it suggests that a member may employ a "scientific justification" *or* Article 5 when, in fact, they have been interpreted as the same.) See also footnote 5 in regard to Article 3.3 cited above. For a statement on the need to examine Article 5 to interpret the basic rights and obligations enumerated in Article 2 see WTO (1998d), which argues that "Articles 2.2 and 5.1 should constantly be read together. Article 2.2 informs Article 5.1: the elements that define the basic obligation set out in Article 2.2 impart meaning to Article 5.1 (para 180)." In addition, the same report (para. 212) notes that Article 2.3 must be read together with Article 5.5—the former declares a general obligation and the latter elaborates "a particular route" for determining whether the general obligation has been met.

These critical provisions in Article 5 essentially yield four rules that countries must follow when they impose SPS measures that deviate from international standards (or when no international standards exist):

(1) The country must obtain a risk assessment (Articles 5.1, 5.2, 5.3, and 5.7).[11]

(2) The SPS measures imposed must be "based on" that risk assessment (Articles 5.1 and 5.7).

(3) The country must not discriminate or create disguised trade barriers by requiring different *levels* of SPS protection in comparable situations (Article 5.5).

(4) The measures must not be more restrictive of trade than necessary to reach the level of SPS protection that the country desires (Article 5.6).

As shown below, the exact meaning of these four requirements is not obvious. However, Article 5 is the linchpin of the SPS Agreement—it puts discipline on SPS protection policies that countries adopt without requiring the politically impossible task of harmonization.

There is a revealing silence in Article 5 and other related provisions of the SPS Agreement.[12] Article 5 is mainly concerned with ensuring that countries base their SPS *measures* on risk assessment and that they not adopt measures that are more restrictive of trade than necessary. It is largely silent on the *level* of SPS protection that a country seeks. Indeed, as mentioned above, several provisions of the SPS Agreement underscore that countries are free to set their own level of SPS protection, even if that level of protection is different from the level that would be afforded by international standards (e.g., Articles 2.1 and 3.3). The only provision in the SPS Agreement that specifically constrains the level of SPS protection that a country may set is Article 5.5, which requires that countries seek comparable levels of SPS protection in comparable situations.[13]

---

[11]The WTO disputes related to risk assessment have focused on Articles 5.1 and 5.2; Article 5.3 is also relevant because it outlines the type of information that should be included in a risk assessment. Article 5.7 concerns provisional measures taken when information is insufficient and is an extension of the basic risk assessment requirements in Articles 5.1, 5.2, and 5.3. In the EC meat hormones case the WTO's Appellate Body noted that Article 5.7 is a reflection of the precautionary principle—in particular, strict measures may be put into place on a temporary basis if information is insufficient (similar statements are found in the sixth paragraph of the preamble and in Article 3.3). However, the precautionary principle and Article 5.7 do not override the requirement to base measures on a risk assessment as denoted in Articles 5.1 and 5.2. See WTO 1998d paras. 120–125. For more on the tests that must be met to qualify under Article 5.7 see the discussion of the Japanese fruits and nuts case below.

[12]The other related provisions are, in particular, Articles 2 and 3 and the definitions in Annex A.

[13]There is a small qualifier to this statement. Article 3.3 also says that members may impose SPS measures "…which result in a higher level of [SPS] protection. . ." *if* one of two conditions is met: the measures are based on a "scientific justification" or the measures are in conformity with Article 5. The concept of "scientific justification" is

Thus, to determine whether a country's level of SPS protection is legitimate one must *look inside the country itself*—at whether the country consistently seeks a particular level of SPS protection. It is possible to interpret the requirements that SPS measures be based on a risk assessment (Articles 5.1, 5.2, 5.3, and 5.7) as also a requirement that a country's SPS levels also be based on risk assessment. Indeed, how can one assess the risks of SPS measures without assessing the risks associated with the level of protection as well? Levels and measures are two sides of the same coin.[14] This remains a hotly contested issue because it concerns perhaps the most politically sensitive aspect of the SPS Agreement— whether it will encroach on a nation's sovereign right to determine its own SPS protection level.

SPS Measures that are Weaker than the International Standard

The other type of exception to harmonization is the reverse of the first: Nations may adopt SPS measures that are *less strict* than international standards. The requirement in Article 5 that standards be based on risk assessment and take into account available scientific evidence applies whether standards are stricter or looser.[15]

So far, none of the formal WTO disputes has addressed SPS measures that are less strict than international standards. Two reasons probably explain why the problem has not arisen: (1) the issue is most prominent in developing countries, many of which are still in transition to full implementation of the SPS Agreement; and (2) for many products, weak SPS measures are much less of a threat to free trade than strong measures. But it is conceivable that this type of exception will come under closer scrutiny and tighter discipline in the future. For manufactured goods, such as processed foods, there is often a substantial premium in efficiency for producers that can export to a market governed by a single standard. Lax standards, even if applied equally to local and imported products, could favor local producers and harm imports that are produced according to more expensive standards that prevail in the rest of the world

---

defined in footnote 5 such that, in practice, scientific justification means based on a risk assessment. The provisions for risk assessment are outlined in Article 5 and in Annex A ("definitions") of the SPS Agreement. Thus the discipline on the *level* of SPS protection that a country may establish funnels through Article 5, and the only part of Article 5 that explicitly addresses the level of SPS protection is Article 5.5.

[14]This is especially evident in the EC's meat hormones ban and Australia's ban on imports of fresh and frozen salmon, where a country's level of SPS protection has been challenged directly. In both cases, the level of protection that the importing country sought was zero risk because the country had imposed a ban on imports. Thus, testing whether the bans were consistent with the requirement to base SPS measures on risk assessment was, de facto, a test of whether the goal of zero risk was based on risk assessment.

[15]The only provision of the SPS Agreement that explicitly applies to national SPS standards that are stricter than international standards is Article 3.3.

market. Using this argument, an alliance of global exporters and environmentalists may discover that the SPS Agreement is a very powerful tool—it could pry open local markets that are "distorted" by weak SPS standards and force a higher level of SPS protection. Whether the SPS Agreement is used in this capacity remains to be seen; such cases probably will be rare, not least because demonstrating the existence of a trade effect is difficult and bringing disputes is costly.

INTERNATIONAL STANDARDS

Although most of the SPS Agreement is focused on exceptions, its principal objective—stated in the preamble—is to promote harmonization of national standards.[16] The SPS Agreement explicitly urges countries to adopt the standards set in three international processes: the *Codex Alimentarius* Commission (food safety), the International Office of Epizootics (animal safety), and the various organizations and processes that operate under the International Plant Protection Convention (plant safety). It also empowers the SPS Committee to identify other appropriate standards, guidelines, and recommendations.

In this section I discuss how these three intergovernmental processes set standards. Most attention is given to the *Codex* process because that has been the most active in actually setting standards and has, by far, attracted the most political attention because the safety of food for human consumption is the most politicized aspect of the SPS Agreement.

The *Codex Alimentarius* Commission[17]

In the aftermath of the World War II, the European nations created several institutions that were designed to promote trade and cooperation. Their architects hoped that the resulting economic integration would widen and deepen—by focusing on making money, Europeans would form a binding political union that would avert future war. The institutions included the European Coal and Steel Community (a predecessor of today's European Union) and the *Codex Alimentarius Europaeus*, established in 1958 to help harmonize methods for testing food safety in Europe. At the same time the

---

[16]Two statements in the preamble make this point: "*Recognizing* the important contribution that international standards, guidelines and recommendations can make in this regard...." and *Desiring* to further the use of harmonized sanitary and phytosanitary measures between Members, on the basis of international standards...." In contrast, the preamble does not mention risk assessment or rules to govern deviations from international standards as principal objectives.

[17]This section is based mainly on Victor (1998). For the early history of *Codex* see Leive, (1976), and Kay, (1976). And for a study with particular attention on pesticide (residue) standards see Boardman, (1986).

World Health Organization (WHO) and the Food and Agriculture Organization (FAO), spurred by the European dairy industry, created a committee to harmonize milk standards and thus open trade in milk and milk products. In 1962 WHO and FAO loosely merged these activities into the *Codex Alimentarius* Commission.

The commission's mandate was to develop and adopt food standards that would allow firms and countries to realize their self-interest: world trade in safe food products. From the outset the emphasis was on participation and consultation, especially with industry; engagement, the *Codex* architects hoped, would lead these stakeholders to harmonize their activities without the need for international enforcement (which was anyway not an available option). Thus, *Codex* standards are developed by committees of government representatives and stakeholders through an eight-step cycle shown in Figure 6-1. Technical committees evaluate evidence and elaborate standards, which are then subjected to the approval of the full *Codex Alimentarius* Commission, which meets every two years. That process of elaboration and approval typically occurs twice (steps 3–5 and 6–8 are a spin cycle), with the goal of ensuring wide input and consensus. Participation in the committee and commission meetings has been open to any stakeholder; yet only rarely have consumer and other public interest groups attended the committee meetings where standards are elaborated. The process is driven by industry, and the vast majority of *Codex* standards attract essentially no attention from other interest groups.

The commission adopts three types of standards: (1) commodity standards, which define what qualifies as a particular commodity (e.g., what is a "canned peach" or "natural mineral water"); (2) residue standards, which define acceptable levels of pesticides and food additives; and (3) codes of conduct and other guidelines that recommend, for example, good practices in the use of veterinary drugs or methods for risk assessment. To date the commission has adopted about 3,000 standards. Here I briefly review three aspects of those standards—how they are created, the role of risk assessment, and the sources of expert advice that are needed to weigh risks. I focus on commodity and residue standards. The other type of *Codex* norm—codes of conduct and guidelines— have been intended to augment application of the core standards rather than as principal standards themselves. In some cases, these looser guidelines have been adopted when agreement was not possible on a commodity or residue standard. However, if the SPS Agreement is interpreted broadly then these looser norms will have potentially binding application—that matter of legal interpretation has not been resolved or tested in any WTO disputes.[18]

The process of setting commodity standards has given practically no attention to risk assessment because most of the work of the *Codex* commodity committees focuses on the physical attributes of the commodity that, indirectly,

---

[18]See footnote 6.

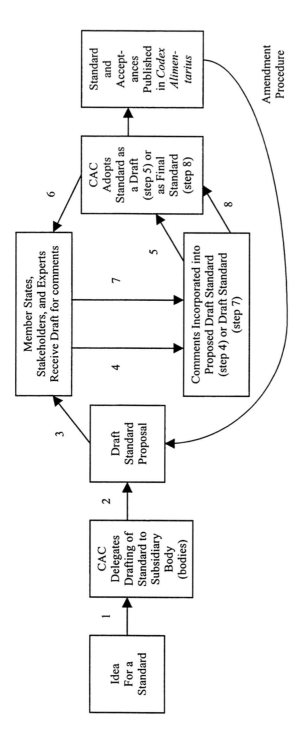

FIGURE 6-1. Elaboration of Food Safety Standards and Other Guidelines by the *Codex Alimentarius* Commission (CAC) and its Subsidiary Bodies. Major milestones (in boxes) and the eight steps by which standards are proposed (steps 1, 2), proposed drafts are reviewed and revised (steps 3,4), adopted (step 5), reviewed and revised again (steps 6, 7), and finally adopted as a *Codex* standard (step 8). Following adoption at step 8, member governments are requested to indicate whether they will accept the new standard. The standards and acceptances are published together as the *Codex Alimentarius*. A similar process is followed for the elaboration of advisory texts, such as codes of practice, except that they are not open for formal acceptance. In cases where the need for a standard is urgent or only one round of review and comments by member governments and stakeholders will be necessary, the commission can choose to omit steps 4–6 and operate an accelerated procedure. The commission can also amend an existing standard, typically sending it back to step 2. Progression throughout the steps is not necessarily linear; the commission often returns a standard to a previous step (e.g., to allow more time for comments and revision).

determine food risks; moreover, there are no specific *Codex* procedures for setting an "acceptable level of risk." Rather, commodity standards are intended to codify what is considered to be good practice for supplying safe food. Thus, de facto, risk assessment—where it exists—enters the *Codex* commodity standards from the "bottom up" through existing industry practice and standards. The committee members themselves provide the needed expertise— committees are populated mainly by government regulators and industry representatives who are best able to define characteristics of a safe canned plum or frozen pea. In practice, this organic and decentralized process led to haphazard commodity standards. Some commodity standards included excessive detail about the attributes of foods that were not necessary for food safety and, instead, merely entrenched existing industrial practices. To remedy this problem, a major review and revamping of *Codex* commodity standards is under way. The goal of that review is to simplify the standards and focus them on safety-related attributes of food products. However, the revamping is not intended to determine particular risk levels or risk assessment procedures that would govern the standard-setting process. So far, none of the WTO disputes related to SPS measures has involved a *Codex* commodity standard.

Many observers have raised the fear that commodity standards are vulnerable to "regulatory capture"—standards set to benefit the standard setters rather than the public interest. Indeed, the *Codex* history gives several suspicious examples. The standards for bee honey effectively barred many non-European honeys from the European market, although there was little basis for doing so on grounds of food safety alone. The worldwide standard for natural mineral water, adopted in 1997, requires that natural mineral waters be bottled at the source, which favors European producers who have long done so according to European law, and prohibits the use of antimicrobial agents that could make water safer. It penalizes American, Japanese, and other producers, many of whom truck or pipe their water prior to bottling and often treat it to ensure its safety. Yet there is not much justification in terms of food safety for the requirement. Piping and trucking do not intrinsically yield dirty water.[19] The incorporation of the *Codex* into the WTO gives standards binding force and may increase the danger that commodity standards will be used for industrial promotion and not only for securing food safety. However, the danger has been longstanding, and incorporation into the WTO has brought other changes that

---

[19]Mindful of this argument, the *Codex Alimentarius* Commission adopted the mineral water standard and supported creation of a separate bottled water standard. When both standards are in place, presumably both types of products will be allowed easy entry into markets. There remains a question of whether waters that are "bottled" but do not meet the at-source requirement for "natural mineral water" can be labeled as mineral water, which often affords the seller a price premium. The issue of labeling is one of the next major topics in the *Codex Alimentarius* Commission and the WTO generally. What disciplines will be applied to labels? To what degree must the use of certain labels be backed by risk assessment or other analytical requirements? For adoption of the natural mineral water standard see *Codex Alimentarius* Commission, (1997, paras. 85–95).

reduce that tendency—in particular, because the *Codex* is now applied in world trade, regional *Codex* standards have been eliminated.[20] A coalition in favor of protectionism is easier to organize when participants are restricted to a like-minded region. Indeed, both the bee honey and natural mineral water standards emerged from regional European standards.[21] Moreover, the "equivalence" provisions of the SPS Agreement (Article 4) will in principle allow exporters to circumvent international standards by using commodity standards that yield an equivalent level of SPS protection.

The *Codex* standard-setting processes for residues have made much more extensive use of risk assessment. Unlike commodity standards, which define all of the major attributes of a particular commodity, residue standards are simply a value for an acceptable residue (the "maximum residue limit" or MRL) of a food additive or contaminant for a particular food. The standards are set by identifying an acceptable daily intake (ADI) of the residue or food additive in question. Typically ADIs are established by identifying an animal that best mimics the most dangerous possible human response to the residue or food additive and determining the "no effect" level in that animal. What is meant by "no effect" and how it translates to human effects has not been rigorously defined or quantified. The ADI for humans is set by adjusting for the mass, diet, and lifetime of a typical human being compared with the test animal. (In the case of the bovine growth hormones, which is used as an example here because that WTO case involved a *Codex* residue standard, the typical human is 60–70 kg and the diet is generously assumed to be 500 g of bovine meat per day over an entire lifetime.) The ADI also includes a large safety factor. (In the bovine growth hormone case, the ADIs are 100 times lower than they would be without the safety factor.) A MRL is then calculated that would ensure that the ADI is not exceeded. If guidelines for "good practice" in food production—for example, guidelines for the use of veterinary drugs, which apply to the use of bovine growth hormones—would yield residues that exceed the MRL, then those guidelines are brought into line. In essence, the *Codex* system adjusts both the "good practice" standards that govern how pesticides, drugs, and food additives are applied during food production as well as the residue standards that govern when the food products themselves are considered safe.[22] In the case of bovine growth hormones, one expert testified that the MRLs adopted by the *Codex Alimentarius* Commission would result in a cancer risk of between 0 and

---

[20]There are still some standards that have only regional application because the way that the standard was developed and the risks it addresses required tailoring to regional conditions—for example, the guidelines on street vending of foods explicitly applies to Africa. But, where possible, the post-WTO *Codex* aims to develop world standards. *Codex Alimentarius* Commission, (1997, paras. 73–75).

[21]For more on protectionism and Codex standards see Victor, (1998).

[22]The process also ensures that the MRLs adopted are consistent with testing equipment and practices for food safety inspection so that the standards are relatively easy to implement.

about one in a million;[23] but that was an estimate because the *Codex* system does not have a standard level of risk that guides its standard-setting activities.

Determining ADIs and MRLs is a highly technical process. Experts are needed to review the raw data from scientific studies and to calculate ADIs and MRLs. The *Codex* system has drawn on the recommendations of two joint WHO/FAO committees that are independent of and external to the *Codex* system: the Joint Meeting on Pesticide Residues (JMPR) and the Joint Expert Committee on Food Additives (JECFA). Both provide advice not only to *Codex* but also to many other activities of WHO, FAO and the United Nations system. In the *Codex*, JMPR and JECFA recommendations are used mainly by the three committees that set residue standards (i.e., MRLs): the Committee on Pesticide Residues, the Committee on Food Additives and Contaminants, and the Committee on Residues of Veterinary Drugs in Foods.

For all three types of *Codex* standards the working committees make recommendations, which they forward to the full *Codex Alimentarius* Commission for decision. To speed its work, the commission allows for simple majority voting when adopting a standard.

Prior to 1994—when the WTO agreements were finalized—the mere adoption of a *Codex* standard had no international legal consequences for *Codex* members. Thus it was rare for *Codex* standards to require a vote because a country could simply ignore an unfavorable standard. Indeed, standards were not binding unless the *Codex* member gave its formal "acceptance." The acceptance process allowed countries to pick and choose which standards they wanted to apply rigorously within their nations. For pesticide residue or food additive MRL standards, a country faced a simple binary choice: accept or not. For more complicated commodity standards, countries could accept the standard "with specific deviations," which gave them the opportunity to unilaterally tune the commodity standard to their own local conditions and preferences.

The combination of extensive consultation in standard setting, simple majority decision making, and the acceptance process makes it difficult to assess what impact *Codex* standards have actually had on national food safety standards and trade. The only hard data come from acceptances, which are not impressive. Table 6-1 shows that by 1993—on the eve of incorporation into the WTO—only 12 percent of the *Codex* standards had been accepted. Moreover, the pattern of commodity standard acceptances suggests that international standards followed rather than shaped national standards: in industrialized countries, which typically already had elaborate national commodity standards in place when international *Codex* norms were developed, nearly all acceptances were "with specific deviations."[24] Deviations allowed them to tune international

---

[23]See statements by the experts in "Annex: Transcript of the Joint Meeting with Experts, held on 17–18 February 1997, WTO (1997b, paras.743, 819, 824, and 826).

[24]Furthermore, most of the full acceptances by advanced industrial (Organization for Economic Cooperation and Development [OECD]) nations were notified by the least developed of the OECD members, such as Portugal.

TABLE 6-1. Acceptances of the *Codex Alimentarius* Commodity Standards (163 standards × 138 countries = 22,494 possible acceptances)

Acceptances	Developing Countries (114 in 1993)	OECD Countries (24 in 1993)	Total
Actual acceptances	2,175	559	2,734
Possible acceptances	18,582	3,912	22,494
Acceptance rate	12%	14%	12%
Type of Acceptance			
Full	1,215 (56%)	100 (18%)	1,315
With specific deviations	228 (10%)	252 (45%)	480
Free distribution	732	207	939
TOTAL	2,175 (100%)	559 (100%)	2,734

Source: Compiled by author from 1989 acceptances, Vol. 14 of *Codex Alimentarius* Commission; updated 1991 and 1993.

standards to meet existing local standards; when the needed deviation was large the country could choose simply not to accept the international standard.

Voluntary standards and the acceptance procedure were designed to give states and stakeholders maximum control over which standards they adopted, which, in turn, dampened potential conflicts. Today, after the incorporation of *Codex* into the WTO, standards are no longer viewed as completely voluntary. Moreover, for purposes of the SPS Agreement, a standard is now considered "adopted" when it has been approved by the *Codex Alimentarius* Commission. The requirement of acceptance, which previously was the way that countries ensured that no *Codex* standard would be imposed against its wishes, no longer plays a role. For example, in the meat hormones case the *Codex* Commission had adopted standards for five of the six hormones in the dispute.[25] The EC did not accept the *Codex* MRL standards, but that nonacceptance was irrelevant to the requirement in the SPS Agreement that the EC base its national standards on international standards (and to provide justification where it did not).

Because of majority voting rules, in principle, the result may be a large number of standards adopted against a country's wishes. Moreover, the large number of *Codex* advisory texts and guidelines now also potentially have binding application through the SPS Agreement. What began as a voluntary body has been transformed into a very different purpose. Conflicts that should

[25] For these natural hormones no MRL was adopted.

have affected the standard-setting process—such as different views on the acceptable level of risk for products, food additives, and residues of veterinary drugs and pesticides—were latent in the *Codex* system but have now developed fully. Indeed, in recent years—especially in the two commission sessions that have been held since the SPS Agreement entered into force (1995 and 1997)—the commission's work is increasingly mired in controversy because it is now viewed as more relevant.

The International Office of Epizootics

The Office International des Epizooties (OIE) is an intergovernmental body established in 1924 with the purpose of protecting animal health. It serves as the umbrella for numerous commissions that prepare codes, protection strategies, and manuals. Some commissions work on specific diseases (e.g., fish diseases or foot-and-mouth disease), others work on problems of specific geographical regions. The OIE periodically revises the *International Animal Health Code (OIE, 1998)* which applies to mammals, birds, and bees; it is also the model for a separate *International Aquatic Animal Health Code* (OIE, 1997).

Both codes include the requirement that countries analyze and manage risks of diseases that are transmitted across borders via international trade and give special attention to adopting measures for controlling diseases that have minimum adverse effects on trade. As with the SPS Agreement itself, the codes also require that countries make their risk analysis transparent and be able to justify their import decisions. In short, the codes provide a basis for establishing quarantines and other sanitary measures and for adjusting the severity of the measures according to the economic risks. However, the requirements strictly apply only to diseases listed in each code; the lists are incomplete and thus offer only a starting point—countries are free to identify other diseases and regulate risks associated with them as well.

In addition to the codes, the OIE also produces guidelines for disease testing and surveillance programs and serves as a clearinghouse for current information on particular diseases (e.g., outbreaks). The work of these commissions is approved by the International Committee, the OIE's main decision-making body. The OIE is also the umbrella for numerous other collaborations that help to develop reference standards; various working groups promote debate that could lead to standards in areas such as biotechnology and wildlife. As of March 1998, 151 countries were members of the OIE.

International Plant Protection Convention

The International Plant Protection Convention (IPPC) entered into force in 1952 and was amended in 1979. It is intended to promote international coordination of measures necessary to limit the spread of plant diseases. The IPPC obliges countries to identify, assess, and manage risks to plants, including risks from plant pests that are carried through international trade. "Guidelines for Pest Risk Analysis," developed within the framework of the IPPC, provide detailed information on how to assess and manage pest risks and require that countries develop import restrictions for protecting plant safety in conjunction with a broader plan for risk management.

The IPPC requires nations to create official plant protection organizations that perform inspections, conduct research, and disseminate information. Most countries would have such organizations in place even without the IPPC. As with the SPS Agreement, it requires that countries adopt phytosanitary measures only to the extent necessary for phytosanitary protection. Countries must use the least restrictive trade measures, avoid unnecessary delays during inspection and quarantine, and ensure that phytosanitary measures are transparent.[26] The IPPC probably aids coordination of national plant protection policies—although some of that would occur anyway among those countries that want to coordinate—but it has not engaged in detailed standard setting to the degree of the *Codex Alimentarius* Commission or the OIE.

Summary

Of the three international standard-setting bodies explicitly mentioned in the SPS Agreement, *Codex* has been extremely active in setting standards for particular SPS hazards. The other two—OIE and IPPC—create mainly procedural obligations to conduct risk assessment and adopt SPS measures that are not excessively restrictive of trade, but those obligations are also enshrined in the SPS Agreement. All three also codify norms of good practice that include the requirement to base SPS standards on risk assessments. But those norms are quite broad. As I show below, they play little role in the detailed process of deciding whether a nation has complied with the SPS Agreement. Even in the *Codex Alimentarius* Commission—where the long experience in setting standards would suggest also long experience in applying risk assessment in formulating those standards—the actual practice of risk assessment is neither

---

[26]The statements here apply strictly to the 1952 IPPC (with revisions that came into force in 1991). A new revised IPPC was adopted by the FAO conference in 1997, but it has not entered into legal force. The new treaty explicitly aligns the requirements of the IPPC with the SPS Agreement, but in practice that has required few significant deviations from the 1952/1991 IPPC Agreement. One significant revision is that the new treaty will create a Commission on Phytosanitary Measures that can provide a standing body to address issues that arise; that body could be important for fine tuning plant-related SPS issues because such matters will probably be more technical than would be appropriate for handling within the SPS Committee (created by the SPS Agreement). Although the new IPPC is not in effect, guidelines for pest risk analysis—adopted in 1995 in parallel with development of the new treaty—probably do apply, regardless of their legal status, because the SPS Agreement has an expansive requirement to base SPS measures on "international standards, guidelines, and recommendations developed by the relevant international organizations...."

transparent nor codified into institutional procedures. Indeed, the lack of codification is perhaps one reason why agreement has been possible. Risks are assessed and standards are set mainly through a bottom-up process that mirrors the risk-averse practice in advanced industrial nations.

OTHER WTO AGREEMENTS: GATT 1994 AND THE TBT AGREEMENT

Two other WTO agreements are often cited in the debate over how to manage technical barriers to trade and thus might be relevant for managing SPS measures. It is worth pausing to consider these two agreements and to explain why they are not centrally important to how risk assessment and discipline have been applied to SPS measures, which is the topic of this paper. Moreover, it is worth explaining why experience with managing nontariff trade barriers under the SPS Agreement should be applied only with great caution to the management of other technical barriers to trade.

First is the General Agreement on Tariffs and Trade 1994 (GATT 1994), which consists of the original 1947 GATT agreement and some revisions (e.g., new tariff schedules). It defines the basic obligations for members of the world trading regime and allows the Article XX "general exceptions" for various national policy purposes. Among the general exceptions is one for SPS purposes (Article XX(b)). However, all of the exceptions remain poorly elaborated and tested. Furthermore, the SPS Agreement declares that if members' SPS measures conform with the agreement that the measures "shall be presumed to be in accordance with the obligations of the Members under the provisions of GATT 1994 which relate to the use of sanitary or phytosanitary measures, in particular the provisions of Article XX(b)." Thus, what matters when determining the discipline on SPS measures is the SPS Agreement, not Article XX(b) of GATT 1994.

Second is the WTO Agreement on Technical Barriers to Trade (TBT Agreement). The TBT Agreement requires that WTO members not use "technical regulations" to discriminate against products imported from other members.[27] The objectives that technical regulations serve must be based on sound science, including risk assessment. The measures employed must not be more trade restrictive than necessary to achieve the objective. The TBT Agreement also urges that, where possible, technical regulations should be based on international standards. Thus the TBT Agreement addresses the generic problems that are the subject of this paper: ensuring that nontariff trade restrictions are not merely disguised protectionism and promoting the use of risk

---

[27]A technical regulation is "[a document] which lays down product characteristics or their related processes and production methods, including the applicable administrative provisions, with which compliance is mandatory. It may also include or deal exclusively with terminology, symbols, packaging, marking or labeling requirements as they apply to a product, process or production method" (TBT Agreement, Annex 1).

assessment and international standards in establishing those restrictions. But the TBT Agreement explicitly states that it does not apply to SPS measures (Article 1.5).

Thus, in practice, neither the GATT 1994 nor the TBT Agreement plays a significant role in governing SPS measures. Nor does the experience with SPS measures examined in this paper directly answer the critical question for Article XX: What threshold must be passed for a trade barrier to be valid as an exception listed in Article XX? Insofar as there is growing clarity in which SPS measures might be considered as compatible with the WTO, it is due mainly to the SPS Agreement. None of the other Article XX exceptions is governed by such a detailed separate agreement or understanding. How to interpret other Article XX exceptions—for example, protection of public morals (Article XX(a)) or exhaustible resources (Article XX(g))—still remains shrouded in mystery.

Great care must also be taken if lessons from this study are applied to non-SPS technical barriers to trade. The TBT Agreement applies a discipline to all technical barriers that is similar to the discipline imposed by the SPS Agreement. In particular, it requires that WTO members not impose technical regulations that are more restrictive of trade than necessary "to fulfill a legitimate objective"(Article 2.2); it requires members to use international standards "as a basis" for their technical regulations (Article 2.4). It establishes procedures that members must follow when they deviate from international standards (Articles 2.9–2.12, Article 4, and Annex 3). It requires members to make their technical regulations transparent (Articles 2.9–2.12) and urges members to treat other members' technical regulations as equivalent (Article 2.7). Thus, many of the same issues arise in both the SPS and the TBT agreements. However, there are important differences between the agreements—the SPS Agreement is narrower in scope. The TBT establishes a broad "Code of Good Practice" for national standard-setting bodies, whereas the SPS Agreement relies on a mixture of specific obligations for national SPS regulatory bodies as well as numerous international guidelines developed by international standard-setting bodies. The TBT Agreement gives close attention to procedures that governments use when implementing technical regulations, whereas the SPS Agreement gives little explicit attention to implementation. The TBT Agreement also requires central governments to ensure that local governments and nongovernmental organizations comply with the agreement, whereas the SPS Agreement does not explicitly address these different layers of regulatory bodies.

Thus, this study attempts to offer insight into only the SPS-related aspects of trade. Some SPS issues also include TBT elements. For example, the dispute over the EC's ban on bovine growth hormones might be resolved by requiring labels on hormone-laced beef and inspection systems to ensure label accuracy. If so, the validity of the label system might be challenged as a technical barrier to trade that is inconsistent with the TBT Agreement. The present study is addressed only to the SPS-related aspects of such disputes.

THE SYSTEM AT WORK: THREE CASES

A full-blown assessment of how the SPS Agreement has affected the use of SPS measures should focus country by country, and measure by measure. That is impractical. The number of trade measures that could be affected by SPS disciplines is potentially huge. So far, only a small fraction has been subjected to international scrutiny. Many changes to national SPS policies will be time-consuming to implement; yet only four years have passed since the WTO agreements went into effect on 1 January 1995.

Thus, the approach here is to examine the three WTO dispute settlement cases that have concerned SPS measures: the EC's ban on imports of bovine meat produced with growth hormones ("EC meat hormones"),[28] Australia's ban on imports of fresh and frozen salmon ("Australian salmon")[29] and Japan's ban on imports of numerous varieties of fruits and nuts ("Japanese fruits and nuts") (WTO, 1998e, 1999). These cases reveal how the SPS Agreement has been interpreted to date and thus are the most instructive means available for beginning to assess the impact of the SPS Agreement.

Prior to the WTO, the dispute settlement procedure had few teeth and was, in essence, voluntary. Any GATT member could block adoption of a dispute panel report and thus block the formal remedies that might help to achieve compliance with trade rules and resolve the dispute. In practice the system was not completely anarchic, but nonetheless it was severely hobbled. The WTO system is more elaborate, has stronger tools at its disposal, is governed by strict timetables that help keep disputes from dragging out over years, and is less vulnerable to dissent. The WTO's Dispute Settlement Body (DSB) manages the process that begins with consultations and other efforts to resolve the dispute. If they fail then the DSB convenes a panel of three experts to hear the arguments of the parties and third parties, consult experts, interpret the relevant WTO obligations, and issue a report with rulings. Either party may appeal the rulings; three members of the standing seven-person Appellate Body review such appeals and issue a report with final rulings. The DSB must adopt Panel and Appellate Body reports; only a consensus of WTO members may block

---

[28]This is actually two cases—one originating from a U.S. complaint and one from a Canadian complaint. But both were heard by the same panel, employed the same experts, were conducted on parallel decision-making tracks, and had the same outcome. See WTO (1997a,b). Both of these cases were appealed, and the WTO Appellate Body issued a single report on the two measures World Trade Organization, (1988d). Finally, the question of what constituted a "reasonable period of time" during which the EC must bring its measure into line was submitted to binding arbitration, which determined that the EC must comply no later than 13 May 1999 (15 months after 13 February 1998, the date of the adoption of the Appellate Body and Panel Reports by the WTO's Dispute Settlement Body). For the outcome of the arbitration see WTO (1998c).

[29]WTO (1998b). The case was appealed; see WTO (1998a). Citations to the Appellate Body report are in the form of page numbers because paragraph numbering is not accurate in the available (online) version of that report.

adoption. Once the final report is adopted, the offending country must comply within a "reasonable period of time."[30]

Formal disputes are important not only because they often address important trade barriers themselves but also because they create interpretations of the law, focus expectations on how the WTO system will handle possible future disputes, and deter other violations. If disputes demonstrate clear discipline and a credible threat to dismantle trade barriers then countries will be more likely to remove illegitimate SPS measures on their own. Indeed, there is substantial evidence that the extended effect is significant—beyond the three measures that have been the subject of formal disputes, the SPS Agreement has been a "broader catalyst" that has induced some nations to remove illegitimate SPS measures (Roberts, 1998). Moreover, as with any properly functioning enforcement system, well-handled disputes can deter countries from imposing illegitimate SPS measures in the future. These extended and deterrent effects could be extremely important multipliers of the effect of individual disputes, but they are also difficult to assess.

The discussion here presents the basic facts and arguments in the cases.[31] In the subsection "Analysis of the System at Work" below, I suggest the major issues and conclusions that should be drawn when examining the whole system: the SPS Agreement, the international standard-setting bodies, and these three cases.

The cases that have been brought to date, as shown below, are relatively clear violations of the SPS Agreement—thus the proper interpretations of central obligations of the SPS Agreement remain fuzzy. The situation may remain that way for a long time. The "transaction costs" of interpreting the SPS Agreement through cases are extremely high. Complainants, defendants, and third parties must prepare complicated arguments; panel members, WTO secretariat staff, and experts must sift through the evidence; the resulting panel decision typically occupies several hundred singled-spaced printed pages. Thus, the system may be inclined to the handling of winner cases in which the challenging member country is relatively sure it will prevail, or highly symbolic cases in which the challenging member country is politically unable to avoid a dispute.[32]

---

[30]See "Understanding On Rules And Procedures Governing The Settlement Of Disputes," Annex 2 of "Agreement Establishing the World Trade Organization." On the matter of a "reasonable period of time"—which is intended to be typically no longer than 15 months—see WTO (1998c).

[31]The discussion of the cases is purposely simplified. The goal here is not to identify the twists and turns in the legal and technical arguments. Rather, it is to identify the main arguments that proved to be most important in resolving the case and thus are likely to have the strongest value as precedents for future cases. The excerpts are based on analysis of the full Panel and Appellate Body reports (WTO, 1997a, b; 1998a, b, d, e; 1999).

[32]Moreover, these high costs raise many issues that are often termed "principal-agent" problems. The beneficiaries of reducing protectionist trade measures are private firms, employees, dependents, and stockholders. But the cost of mobilizing and

EC Meat Hormones[33]

The first case concerns an EC Directive, imposed in 1981 and strengthened in 1988 and 1996, to ban imports of meat from farm animals that had been administered natural or synthetic hormones. Exceptions were allowed for hormones that are used for therapeutic purposes but not for hormones used to promote growth in cows. American, Canadian, and other beef producers used hormones to accelerate growth while reducing costs and yielding higher quality (leaner) meat. The United States had challenged the EC ban under the Tokyo Round "code" on technical barriers to trade, but the EC had blocked formation of an expert panel to examine the dispute. The conflict festered and became symbolic of why the voluntary Tokyo Round codes and nonmandatory dispute settlement were incapable of imposing discipline on nontariff barriers to trade.

At issue was whether the EC ban, which concerned six hormones, was compatible with the SPS Agreement. In 1995 the *Codex Alimentarius* Commission had adopted standards for five of the six hormones in the dispute. The standards were based on the work of the *Codex* Committee on Veterinary Drugs in Foods and the recommendation of JECFA, which had reviewed the scientific evidence related to hormones twice. The *Codex* standards did not impose MRLs for the three natural hormones in question (oestradiol-17ß, progesterone, and testosterone) because naturally produced residues would far exceed the additional residue caused by "good-practice" use of these hormones for promoting growth in cows. For the other two synthetic hormones (trenbolone acetate and zeranol, which mimic the biological activity of natural hormones), the MRLs adopted were far less strict than the level that would be expected if good veterinary practices were followed. There were no *Codex* standards for melengestrol acetate (MGA), a synthetic hormone administered as a feed additive that was included in the EC ban.

The EC argued that the SPS Agreement explicitly allows WTO members to adopt standards that are stricter than international norms if those standards are based on an assessment of risks. Every risk assessment of these hormones had shown that growth hormones applied according to good veterinary practices would result in no significant harm to humans—those assessments included two major reviews by JECFA (1988 and 1989) and at least two reviews commissioned by the EC itself.[34] The EC argued that, although those studies

---

prosecuting a case—including tasks such as commissioning new scientific studies and risk assessments that bolster the claim—are typically borne by governments. I do not address this issue further here, except to note that if transaction costs remain high, these principal-agent concerns could be especially severe. It is governments that do most of the work in maintaining international legal agreements—including arguing WTO disputes— but the point of globalization and privatization is to empower private actors to seize the benefits of liberal rules.

[33]For more on the origins of this dispute see Vogel (1995, Chapter 5); for more on the WTO aspects of the dispute see Charnovitz (1997), and Roberts (1998).

[34]*32nd JECFA Report*, published in 1988 (1988 JECFA Report); *34th JECFA Report*, published 1989 (1989 JECFA Report); Report of the Scientific Group on

suggested that there was no objective risk, numerous highly publicized incidents since the early 1980s during which hormones entered European food markets had made European consumers wary of beef.[35] A ban, the EC argued, was necessary to restore confidence in the market.[36]

The WTO Dispute Panel ruled against the EC on three grounds. First, it argued that the EC's measure was illegal because more-permissive international standards existed for five of the hormones. The Dispute Panel interpreted Article 3.1 of the SPS Agreement, which declares that "Members shall base their sanitary or phytosanitary measures on international standards" as a requirement that SPS measures *conform* with international standards.[37] In perhaps its single most important ruling on SPS-related issues, the WTO Appellate Body explicitly overturned this interpretation, preferring instead the more common-sense definition of "based on:" A measure can be based on international standards without conforming with those standards. Instead of conformity, the Appellate Body pointed to Article 3's fundamental purpose: to promote the use of international standards while allowing countries to deviate from those standards if those deviations conform with Article 5 which pertains to the use of risk assessment (WTO (1998d), paras. 160–177). This approach of the Appellate Body, although obviously more consistent with the purpose of the SPS Agreement than the narrow interpretation imposed by the Dispute Panel, was nonetheless a watershed—it removed a legal interpretation that could have resulted in international standards becoming the feared straitjacket.

Second, the Dispute Panel also ruled that the EC measure was not based on a risk assessment as required in Article 5.1. The Appellate Body agreed. The Panel and Appellate Body found this for five of the hormones that the EC had obtained assessments of some risks. Among these assessments, only a 1982 report of the EC Scientific Veterinary Committee (the Lamming Report) and two reports by JECFA (1988 and 1989) qualified as adequate risk assessments.[38] The Appellate Body

---

Anabolic Agents, Interim Report, 22 September 1982 (Lamming Report); EC Scientific Conference on Growth Promotion in Meat Production, 29 November to 1 December 1995 (1995 EC Scientific Conference). For a conclusion from the 1995 EC Scientific Conference that starkly states that growth hormones are safe, see Maddox (1995).

[35]The EC did cite some risk assessments that pointed to a risk of cancer due, broadly, to hormone exposure. However, those assessments did not examine the risks associated with particular hormones and were not treated as relevant evidence by the Panel, especially as numerous other more-focused assessments showed no particular risk.

[36]For the arguments, including quotes from European Parliament reports favoring a ban, see World Trade Organization (1997b, paras. 2.26–2.33).

[37]In particular, the panel decided that "based on" meant that the SPS measure should afford the same level of SPS protection as the international standard. See WTO (1997b), para. 8.72.

[38]Other reports were also presented by the EC and other members as "risk assessments" but they were discounted. Some were cursory examinations of the issues. In particular, the EC's strongest evidence that hormones caused risks were in reports that examined only categories of hormones or the hormones at issue in general. Those studies were discounted as not adequately focused. See WTO (1998d, paras. 195–202).

underscored that risk assessments need not be based entirely on research in the physical sciences; nor must risk assessments examine only quantitative risks. However, the EC measure failed because the EC had not applied risk assessment techniques to the particular risks that the EC claimed were the basis of its SPS measures (an import ban). The EC had argued, for example, that misuse of hormones as growth promoters could cause excessive risks and thus all use of hormones for growth promotion must be banned, but the Appellate Body concluded that the EC had not actually presented an assessment of such risks (WTO, 1998d, paras. 207–208). Hence the conclusion that the EC measures were not based on a risk assessment. Moreover, the Appellate Body decided that not only is there a procedural requirement to obtain a risk assessment, but in addition: "The requirement that an SPS measure be 'based on' a risk assessment is a substantive requirement that there be a *rational relationship* between the measure and the risk assessment" (emphasis added).[39] The fact that all of the valid risk assessments showed that "good practice" application of growth hormones was safe—and the failure to examine the risks that the EC claimed could result in harm to consumers—meant that the EC measure failed the "rational relationship" test. But the exact contours of that test remain unexplained.

For the other hormone (MGA), no valid risk assessment existed and thus, by definition, the EC measure was not "based on" a risk assessment (WTO, 1998d, para. 201).[40]

Third, the Panel found that the EC had violated Article 5.5 of the SPS Agreement by demanding different levels of SPS protection in comparable situations. Notably, the EC allowed carbadox and olaquindox to be used as antimicrobial feed additives that promoted the growth of pigs; yet the EC banned the use of hormones as growth promoters in cows although the hormones resulted in similar (or lower) risks to humans. The Appellate Body overturned that decision by declaring that the SPS level required by a country would be incompatible with Article 5.5 if it failed *each* of the following three tests: (1) the country did not require comparable levels of protection in comparable situations, (2) the failure to apply comparable measures in comparable situations is arbitrary and unjustifiable, and (3) such measures result

---

[39]For this quote and the elliptical endorsement of the panel's approach, see WTO (1998d, para. 193).

[40]Due to the lack of evidence, the EC might have maintained the ban on MGA as a "provisional" measure under Article 5.7 of the SPS Agreement. However, the WTO Dispute Panel dismissed that argument because the EC did not claim that the measure was "provisional" and concluded that the ban on MGA still would need to comply with the other provisions of the SPS Agreement (e.g., the requirement to conduct a risk assessment). See WTO (1997b, para 8.248–8.249 and paras. 8.250–8.271. The EC might have overturned at least part of that ruling on appeal which could have, perhaps, allowed the MGA ban to stand under Article 5.7's allowance for strict measures in the face of uncertainty (in essence, the "precautionary principle"). However, this was not a central issue in the appeal and the Appellate Body did not rule on that particular argument (i.e., Article 5.7) directly; and generally the Appellate Body did not view the precautionary principle as giving countries wide latitude (see footnote 11).

in discrimination or a disguised restriction on international trade.[41] The Appellate Body found that the EC had, indeed, applied different SPS levels in comparable situations and thus failed the first test.[42] The EC ban also failed the second test because the EC could not justify this difference in treatment. But the Appellate Body argued that the third test—whether "arbitrary or unjustifiable" differences in SPS levels harmed trade—was most important, and the complainants provided insufficient evidence that the EC measure failed that test. Allowing carbadox and olaquindox as feed additives on the one hand while barring hormones for promoting growth in cows on the other was not by itself evidence of a disguised barrier to trade. Erecting a trade barrier was not the purpose of the EC rules that created this incongruous situation—in the words of the Appellate Body the "architecture and structure" of the EC Directives was not discriminatory or a disguised restriction on trade. The EC applied the same level of SPS protection (with a ban on hormones as growth promoters) equally to imports and domestic production. Nor had the United States or Canada submitted adequate evidence that the different treatment had resulted in "discrimination or a disguised restriction on international trade."[43]

In sum, the Panel viewed the SPS Agreement as requiring strict adherence to international standards and sharply limiting a nation's right to determine its SPS levels and measures. The Appellate Body, which is more attuned to the political and social context in which the SPS Agreement and the WTO operate, gave importers much greater autonomy in setting SPS policy. Whereas the Panel found three main reasons to rule against the EC, the Appellate Body endorsed only one—the EC's failure to base its SPS measures on a risk assessment.[44]

---

[41]The Appellate Body derived this three-part test in part from Article 5.5, which requires that "each Member shall avoid arbitrary or unjustifiable distinctions in the levels [of SPS protection] it considers to be appropriate in different situations." The interpretation of that requirement requires, in part, looking to Article 2.3 of the SPS Agreement which is part of the agreement's basic rights and obligations: "Members shall ensure that their sanitary and phytosanitary measures do not arbitrarily or unjustifiably discriminate between Members where identical or similar conditions prevail, *including between their own territory* ..." (emphasis added). For the three-part test see WTO (1998d, paras. 210–246).

[42]In addition to allowing the use of carbadox and olaquindox while banning growth hormones in beef, the WTO Panel had also suggested that there were many other examples where the EC had not applied comparable levels of protection in comparable situations. The panel drew particular attention to the fact that the natural residues of these hormones were higher in some foods—such as eggs and broccoli—than would occur if applied as growth promoters. The Appellate Body rejected these comparisons because the addition of hormones for growth promotion was different from the natural presence of hormones in food—the former concerns an intervention by humans in the food production process, whereas the latter is a fact of nature that humans cannot alter without a "comprehensive and massive governmental intervention in nature." See WTO (1998d, para. 221).

[43]For the third part of the test, see WTO (1998d, paras. 236–246).

[44]Of course the dispute also touched on many other issues—here I have raised only the most important ones that related directly to the interpretation of the SPS Agreement

Australian Salmon

This dispute, the second involving SPS measures to result in a Panel decision, concerned an Australian regulation dating from 1975 that bans imports of fresh or frozen salmon in order to prevent 24 fishborne diseases from spreading into Australia's pristine environment. Many of the diseases could adversely affect trout, which are vital to Australian sport fishing and tourism as well as a small trout aquaculture industry. And the diseases could also harm the Atlantic salmon aquaculture farms, first established in 1986 in Tasmania, that export salmon to world markets and also sell their product on the local Australian market. To combat the threat, Australia required heat treatment for all imports from regions where fish might become infected with the diseases.

The OIE listed two of these diseases in the *International Aquatic Animal Health Code* category of fish diseases that are particularly dangerous threats for spreading. Such transmissible diseases "are considered to be of socio-economic and/or public health importance within countries and that are significant in the international trade of aquatic animals and aquatic animal products" (OIE, 1997, Section 1.1). The OIE also listed four of the diseases in a category of fish diseases that are less well understood but potentially dangerous. For diseases on either list, OIE "Guidelines for Risk Assessment" require countries to undertake analysis to examine the "disease risks associated with the importation" and to tailor particular import controls to the real-world situations in the country.[45] The remaining diseases were not listed by OIE and thus no special OIE guidelines were applicable.[46]

---

and the effect of the SPS Agreement on nations' SPS policies. Among the other issues is the burden of proof. The Panel argued that the importing (defending) country had the obligation to prove the consistency of its SPS levels. The Appellate Body argued that the complainant must first establish a prima facie case that the defending country violated the SPS Agreement; only then must the defender disprove the claim. The Appellate Body also addressed procedural issues related to the handling of matters related to the WTO's dispute settlement procedures and whether a dispute could be prosecuted for measures, including the EC hormone ban, that were imposed before 1 January 1995 (the date when the WTO agreements came into force).

[45]The Guidelines are codified in the OIE (1997, Sections 1.4.2.1–1.4.2.3).

[46]The *International Aquatic Animal Health Code* does include a more general requirement that countries conduct "import risk analysis to provide importing countries with an objective and defensible method of assessing the disease risks associated with the importation of aquatic animals, aquatic animal products, aquatic animal genetic material, feedstuffs, biological products and pathological material" (Section 1.4.1.1). A liberal interpretation of the Code would suggest that that requirement applies generally to imports and not only to listed diseases. However, the Code explicitly allows countries to determine their own methodology for conducting such analysis; countries can use procedures outlined in OIE reference documents for conducting such analysis, but they are not required to do so (Section 1.4.1.3). Moreover, the broad requirement to conduct import risk analysis also exists in the SPS Agreement. Finally, the definition of "disease" in the *International Aquatic Animal Health Code* strictly applies only to diseases that are included on one of the Codes two lists.

Canada, a major exporter of fresh and frozen salmon, challenged Australia's regulation. Canada did not dispute that Australia had the right to preserve a pristine environment—that is, in the jargon of the SPS Agreement, Australia had the right to determine its own "appropriate level of SPS protection." But, Canada argued, the quarantine was arbitrary because Australia did not apply similarly strict quarantine measures against other disease risks. Australia had allowed imports of frozen herring bait fish and live ornamental fish that could much more easily transmit many of the 24 diseases into Australian waters, but it barred Canadian salmon. Bait fish are, by design, disposed directly into waters where diseases could easily pass to other fish. Ornamental fish often escape their ponds and aquaria; when they die they may be disposed without care for the risk of transmitting diseases to other fish in Australian waters. In contrast, headless and eviscerated fresh or frozen salmon from Canada had low incidence of the diseases and could transmit the disease into the Australian fish population only through a long and implausible chain of events.[47] Numerous risk assessments supported the Canadian argument. As the EC argued in the meat hormones case, Australia maintained that, although the risks were low, it could not be certain that headless eviscerated fish would not spread disease.

The Panel and Appellate Body ruled against the Australian measure largely on three grounds. First, the Appellate Body determined that Australia's ban on imports of fresh and frozen Canadian salmon was not based on an assessment of risks. In doing so, the Appellate Body established a three-pronged test for what would qualify as a risk assessment: (1) identification of the diseases and possible biological and economic consequences of their entry or spreading; (2) evaluation of the likelihood of entry, establishment, or spreading; and (3) evaluation of the impact of SPS measures on the likelihood of entry, establishment, or spreading of the diseases.[48] Australia's "1996 Final Report," which established the ban on imports of fresh and frozen salmon, met the first requirement. But the Appellate Body said that Australia had failed the other two tests. This finding overturned the Panel, which had ruled that the 1996 Final Report did constitute a "risk assessment." The Panel had followed the cue of the earlier Appellate Body report on EC meat hormones, which suggested that the requirement of the SPS Agreement be "based on an assessment," which implied

---

[47]An example of the chain of events required: A disease-ridden fish carcass would be disposed of in the sewers, sewage would leak into waterways, and waterways would then carry the disease (perhaps via an intermediate host) into the Australian fisheries. Canada argued that the probability of each step was low and, in total, the probability of the full chain of events was extremely low. The case focused on Pacific wild salmon, which were the most important potential Canadian export and had been the subject of a special effort by Canada and the United States to perform a risk assessment and obtain export permission from Australia. Later that same risk assessment process would be extended to other species. Such risk assessments must differentiate between populations and species because the incidence of disease and risk of transmission probably vary.

[48]The three-pronged test is based on Article 5.1 and Annex A (para. 4) of the SPS Agreement. For the test, see WTO (1998b, p. 73).

that WTO members could include many diverse factors. But the Panel had wrongly assumed that this permissive standard also meant a low threshold for what qualified as a risk assessment. The Panel concluded that the 1996 Final Report "to some extent evaluates" the risks and risk reduction factors and thus qualifies as a risk assessment, but the Appellate Body established a stronger test for compliance.

Second, the Panel and Appellate Body found that the import ban on fresh and frozen salmon was a disguised restriction on trade. Both the Panel and the Appellate Body stressed that Australia was free to determine its own level of SPS protection; however, they found that Australia did not apply that high level of protection in other comparable situations. By allowing imports of bait and ornamental fish, Australia exposed itself to greater risks than from salmon imports; not treating these comparable risks in comparable ways revealed that the salmon import ban was a disguised restriction on trade. To reach this decision the Panel applied the three-step test that the Appellate Body had developed in the EC meat hormones case: (1) it decided that the situation of disease risks from salmon imports was comparable with the disease risks from ornamental and bait fish because they involved similar diseases, media, and modes of propagation; (2) such different treatment for salmon and other disease risks was "arbitrary or unjustifiable;" and (3) the different treatment for salmon resulted in a disguised restriction on international trade. The Appellate Body agreed. Whereas the third element of the test failed in the EC meat hormones cases, the evidence was much stronger in the salmon case. The evidence included the fact that the draft of Australia's salmon rules would have permitted the importation of ocean-caught Pacific salmon under certain conditions; but the final rule—issued after stakeholders such as the Australian salmon industry had commented but based on substantially the same risk assessment information— barred imports. That factor, compounded by many other "warning signals," led the Panel and Appellate Body to decide that the import ban was, indeed, a disguised restriction on trade.[49]

Third, the panel decided that the particular SPS measure required by Australia—heat treatment of salmon prior to export to Australia—was more trade restrictive than necessary and thus violated Article 5.6 of the SPS Agreement. Heat treatment, in effect, barred Canadian salmon from a lucrative segment of the market because heat treatment, by definition, converted fresh or fresh-frozen fish into less valuable heat-treated fish. (Moreover, some experts consulted by the Panel suggested that heat treatment might actually raise the

---

[49]The Panel's ruling on all the major issues in this case was developed by focusing on ocean-caught Pacific salmon because those were the first that Canada sought to export. However, similar issues arose for other salmon because the import ban applied to all Canadian fresh and frozen salmon, and where possible the Appellate Body extended its ruling to cover other salmon as well. (Salmon stocks must be considered separately because some of the disease risks vary with the ecosystem in which the salmon are caught.) For the three-part test applied to ocean-caught Pacific salmon, see World Trade Organization (1998b, pp. 80–93). For the test applied to other salmon see pp. 108–111.

disease risks because elevated temperatures were not high enough to kill all pathogens and could cause some to grow more rapidly.) An alternative sanitary measure—requiring the beheading and evisceration of fish—would yield a similar level of SPS protection for Australia with a much less deleterious impact on Canada's exports. The Appellate Body appeared to be inclined to agree with the Panel, but it overturned this aspect of the ruling. The Appellate Body argued that the SPS measure at issue was not heat treatment but rather the import ban on fresh and frozen salmon from Canada. (Because of that ban, the only means available to Canada to supply salmon to the Australian market was heat treatment.) The Appellate Body overturned the Panel because it could not determine Australia's "appropriate level of protection" and, therefore, could not determine whether the Australian measures were excessively restrictive on trade. The Appellate Body underscored that "determination of the appropriate level of protection ...as a prerogative of the Member concerned [Australia]...."[50]

Japanese Fruits and Nuts

The final case concerns a Japanese regulation that had the effect of requiring exporters of various fruits and nuts to submit each new variety they intended to export to Japan to an extensive regime to verify that fumigation with methyl bromide would effectively kill the eggs and larvae of coddling moths.[51] The case focused on four species (apples, cherries, nectarines, and walnuts), although it potentially had application to others. The required treatment varied not only with the characteristics of the fruit or nut but also the season of harvest because coddling moths exist in different forms (e.g., eggs, larvae, adults) in different seasons. Different varieties have different harvest times, and thus Japan argued that test results for one variety were not applicable to another.[52] The United States challenged the requirement as not based on an assessment of risks; it also argued that the varietal testing requirement imposed excessive costs and delays and thus was more trade restrictive than required. The United States

---

[50]The ambiguity reflects that Australia's measure (the import ban) was not based on a risk assessment—in particular, it failed to assess the risk reduction that might be caused by alternative SPS measures. Australia maintained that its level of protection was "very conservative" (Panel Report, para. 8.107); but its prohibition on imports suggested that the actual level of SPS protection that Australia sought was zero risk. On ocean-caught Pacific salmon, see World Trade Organization (1998b, pp. 93–104); for other salmon see p. 112. For the quotation here see p. 99.

[51]The case also included attention to nonfumigation techniques (cold treatment).

[52]The United States challenged the Japanese varietal testing requirement for all "US products on which Japan claims that coddling moth may occur," which included apricots, pears, plums and quince. But the United States had not provided a prima facie case that the Japanese testing requirement was maintained "without sufficient scientific evidence." The United States met that standard for apples, cherries, nectarines, and walnuts but not for the other four fruits. See WTO (1999 paras. 132–138).

contested only the measures that Japan had applied; it explicitly did not question Japan's right to determine its "appropriate level of SPS protection"—that is, for Japan to ensure that its pristine islands remain free of the coddling moth (WTO, 1998e, para. 827).[53]

The Panel found that Japan's testing requirements were inconsistent with the SPS Agreement for three reasons. First, the varietal testing requirement was not based on a risk assessment. (The failure to employ risk assessment also violated the IPPC's requirement to base plant protection measures on risk assessments. However, in practice, the IPPC's requirements were redundant of the SPS Agreement's obligation to base measures on risk assessment; thus the IPPC played no significant role in this dispute.) In particular, the Panel concluded that "it has not been sufficiently demonstrated that there is a rational or objective relationship between the varietal testing requirement and the scientific evidence submitted to the Panel" (WTO, 1998e, para. 827). Japan claimed that its goal was to ensure that new varieties would impose no danger of coddling moth infestation that was greater than the infinitesimal risk of infestation from varieties that had already undergone extensive testing. Each variety must be tested individually, Japan argued, because there may be a chance (although extremely small) that differences between varieties of fruits and nuts could lead to ineffective treatments that would let a coddling moth slip through. However, the Panel found that " ...so far not a single instance has occurred in Japan or any other country, where the treatment approved for one variety of a product has had to be modified to ensure an effective treatment for another variety of the same product (WTO, 1998e, para. 872)." Moreover, the United States as well as experts advising the Panel had shown that varietal differences did not influence the efficacy of quarantine methods, and Japan had not presented adequate evidence to the contrary (WTO, 1998e, para. 827).[54]

The Appellate Body endorsed the conclusion that the Japanese testing requirement was not based on a risk assessment; echoing Article 2.2 of the SPS Agreement, the Appellate Body found that the testing requirement was maintained "without sufficient scientific evidence" (WTO, 1999, para. 76, and SPS Agreement Article 2.2). However, as in the hormones and salmon cases, the

---

[53]Ensuring that Japan would remain free of the coddling moth is, of course, impossible to guarantee. Japan's requirement is that all 30,000 insects at the most resistant stage in their development die in large-scale fumigation tests. Japan considers that efficacy as equivalent to at least a 99.9968 percent ("probit 9") treatment efficacy. See WTO (1998e, paras. 2.15 and 2.23). In addition to this large-scale mortality test, there are preliminary ("basic") small-scale tests and on-site confirmatory tests. The Japanese varietal testing requirement obliged exporters to perform the basic test and on-site confirmatory tests for each variety, but the large-scale mortality test need not be repeated for each variety. See para. 2.23 and 2.24.

[54]Data did exist to show that the measurements that are typically used to determine quarantine efficiency varied across tests on different varieties. However, the United States argued (and experts advising the Panel confirmed) that the differences were easily due to differences in testing conditions and did not indicate substantive differences in the efficacy of the varietal testing requirement.

Appellate Body also avoided creating any standard for "sufficient" or "rational relationship;" instead, they found "[w]hether there is a rational relationship between an SPS measure and the scientific evidence is to be determined on a case-by-case basis and will depend upon the particular circumstances of the case, including the characteristics of the measure at issue and the quality and quantity of the scientific evidence" (WTO, 1999, para. 84).

Japan argued that Article 5.7 allowed countries to adopt stringent measures when "relevant scientific evidence is insufficient." The Panel underscored that Article 5.7 is an exception to the general risk assessment obligations of the SPS Agreement (i.e., Articles 2.2 and 5.1) that applies only to *provisional* measures. The language of Article 5.7 itself suggests that such provisional measures must meet four cumulative requirements:

(1) the measure is imposed where "relevant scientific information is insufficient,"

(2) the measure is adopted "on the basis of available pertinent information,"

(3) the member must "seek to obtain the additional information necessary for a more

objective assessment of risk," and

(4) the member must "review the ...phytosanitary measure accordingly" within a reasonable period of time" (SPS Agreement, Article 5.7).

The Panel concluded that Japan had failed on at least both the third and the fourth requirements (WTO, 1998e, paras. 8.49–8.60).

Second, the Panel also found that the varietal testing requirement was more trade restrictive than necessary and thus violated Article 5.6 of the SPS Agreement. Because there is no significant difference in the efficacy of fumigation techniques across different varieties of the same product, alternative measures—such as setting fumigation requirements on the basis of the easily measured "sorption level" of new varieties, rather than a full retesting of each variety—would be less restrictive of trade yet still achieve the level of SPS protection that Japan requires (WTO, 1998e, paras. 8.70-8.104).[55] The Appellate Body overturned this ruling because it was based on evidence marshaled by the Panel itself and thus the Panel had overstepped its authority;[56] the United States had not, first, presented a prima facie case that a measure based on determination of sorption levels would have met the requirements of Article 5.6 and thus been less trade restrictive than the Japanese varietal testing scheme (WTO, 1999, paras. 123–131).[57]

---

[55]The Appellate Body agreed, see WTO (1999, paras, 86–94).

[56]The idea for a "determination of sorption level" approach derived from suggestions from the experts advising the Panel (see Panel Report, para. 8.74).

[57]In addition to the "determination of sorption levels" approach, the Panel also considered another measure in-depth, which the United States had proposed as valid under Article 5.6: testing by *product* rather than *variety*. The Panel rejected that

Finally, the Panel and Appellate Body found that Japan had violated the requirement to make its SPS measures transparent, especially the requirement in Article 7 that countries publish their SPS measures. The Japanese varietal testing requirement was based on numerous de facto rules that were not easily understood by outsiders, which made it difficult for exporters to understand and comply with the requirements of the Japanese market.

ANALYSIS OF THE SYSTEM AT WORK

Here, I examine the legal requirements of the SPS Agreement and the first three SPS cases from the perspective of several major conclusions that can be drawn at this early stage of implementation.

First, however, it is important to underscore that the legal interpretation of the agreement is still evolving. The SPS Agreement was written carefully to allow countries to determine their own "appropriate level" of SPS protection and to determine which SPS measures are necessary to meet that level. It would not have been politically possible to adopt an SPS Agreement that forced strict harmonization of either the level or measures for SPS protection. Nor would strict harmonization be necessary to avoid most of the trade distortions caused by SPS measures. The agreement thus aims to promote trade by imposing strict disciplines on the process by which members set SPS protection levels and measures that affect international trade. Judging the success of the SPS Agreement thus requires determining whether and how those disciplines have been implemented, rather than whether particular world standards have been adopted or even whether nations have adopted the same SPS levels and measures.

The three WTO disputes have helped to clarify the obligations in the SPS Agreement, but the disputes have also left many areas still uncertain. One, perhaps the most important, is the central obligation of the SPS Agreement: SPS measures must be "based on scientific principles," in particular an "assessment ...of the risks" (Articles 2.2 and 5.1, respectively). All three cases underscore that there is a difference between *risk assessment* and *risk management*—the former is a scientific process that examines the magnitude and distribution of possible risks, and the latter employs risk assessment as well as many other factors in determining and attaining the appropriate level of risk. The hormones case underscores that the SPS Agreement does not mandate a particular quantified relationship between risk assessment and risk management. Indeed, the Appellate Body explicitly underscored that when WTO members

---

approach because "it is not possible to state with an appropriate degree of certainty that one and the same treatment would be effective for all varieties of a product" (WTO, 1998e, para.8.83). The Panel also considered two variants on "sorption level" approaches. One involved monitoring the "c x t value" (average fumigant gas concentration multiplied by the fumigation). The Panel found insufficient evidence to determine the efficacy of that approach. The other is the "determination of sorption level" approach discussed in the text. See WTO (1998e, paras. 8.85–8.104).

determine their SPS measures (i.e., when they manage risks) that many nonquantitative factors (e.g., consumer confidence) may enter the equation.[58] But in the hormones case the SPS measures were ruled invalid because the EC appeared to have made *no use* of risk assessment when determining its SPS measures. Indeed, in that case and the other two cases, WTO panels or the Appellate Body showed that risk assessments conducted by the offending governments themselves did not support the imposition of such extreme trade controls.[59] In all three cases the defendants nonetheless supported their actions because, they argued, there *could be* some risk if the imported products were not subjected to stringent SPS measures. The WTO panels and Appellate Body rejected that argument because science can never prove that an action will result in absolutely zero risk, and thus policy-making requires the assessment of risks. The Appellate Body in the meat hormones case set the standard which has since prevailed: There must be a "rational relationship" between risk assessment and a country's SPS measures.[60] But the Appellate Body defined neither "rational" nor "relationship." The standard is both procedural and substantive, but the more important substantive element is unclear. By allowing many nonquantitative factors to enter the equation, the Appellate Body appears to have set a low threshold for meeting this standard, but how low is it? In the Japanese fruits and nuts case the Appellate Body shed no further light when it underscored that this determination must be made on a case-by-case basis (World Trade Organization, 1999, and text above).

Second, it is also unclear how to interpret the requirement in the SPS Agreement which bars countries from setting SPS protection levels that "arbitrarily or unjustifiably discriminate between Members where identical or

---

[58]For example, the Appellate Body report in the salmon case (WTO, 1998d, para. 194), underscores that risk assessment need not "come to a monolithic conclusion that coincides with the scientific conclusion or view implicit in the SPS measure. The risk assessment could set out both the prevailing view representing the 'mainstream' of scientific opinion, as well as the opinions of scientists taking a divergent view." The Appellate Body also concluded that "the scope of a risk assessment [can include] factors which are not susceptible of quantitative analysis." (WTO, 1998e, para. 253(j)). See also para. 245, where the Appellate Board explicitly endorsed consumer fears and other factors that can be included as legitimate when a WTO member decides which SPS measure to apply. Underscoring that a wide range of factors may be included in a risk assessment and in the development of SPS measures "based on" a risk assessment, see the Appellate Board report from the Australian salmon case (WTO, 1998a, p. 74).

[59]In the meat hormones case the European Union sponsored two major assessments—neither concluded that there was any significant consumer risk. In the salmon case the Australian government concluded that Canadian imports would not pose a risk and drafted a policy to allow those imports; after a public comment period, the government reversed the policy proposal and banned fresh and frozen imports that had not undergone heat treatment.

[60]Specifically, the Appellate Board declared that the SPS Agreement's requirement that an SPS measure be "based on" a risk assessment is "a substantive requirement that there be a rational relationship between the measure and the risk assessment" (WTO, 1998d, para. 193).

similar conditions prevail..." (Article 2.3). That requirement is echoed in Article 5.5, which requires that countries apply measures that yield a comparable level of protection under comparable situations.[61] The SPS Committee is charged with developing guidelines to distinguish comparable from noncomparable situations, but that will not be easy. In two of the three disputes to date, it has been relatively easy to identify a comparable situation where allowable SPS risks were substantially different. In the EC hormones case, carbadox and olaquindox were allowed for swine production, but natural and synthetic hormones were not allowed for beef production; yet both substances have been linked to a similar health effect (carcinogenicity); some experts consulted by the WTO Dispute Panel even suggested that such feed additives could directly harm workers who handled feeds.[62] In the salmon case, Australia allowed imports of frozen herring bait and live ornamental fish that harbored many of the same diseases that Australia feared would arrive on imported salmon; yet fresh and frozen salmon were effectively barred, which protected the nascent Australian aquaculture industry.[63] These easy cases have not given much clarity to what is comparable, except to underscore that comparable will not be interpreted in the narrowest possible manner as a requirement to impose comparable levels of SPS protection only when exactly the same sources and types of risks are at stake.[64]

A third area of important ambiguity concerns the criteria for judging the trade effects of SPS measures. The SPS Agreement is not a catchall requirement that countries must base all SPS levels and measures on science and risk assessment. Rather, it seeks only to bar SPS levels and measures that cause

---

[61]The exact language, from Article 5.5, is "each Member shall avoid arbitrary or unjustifiable distinctions in the levels it considers to be appropriate in different situations, if such distinctions result in discrimination or a disguised restriction on international trade." But the language of "different situations" is counterintuitive and awkward to use in plain English. Thus, throughout this paper I have followed the approach of the Panel in the EC meat hormones case and use the term "comparable situations."

[62]The argument that carbadox and olaquindox were "comparable situations" and thus proof that the European Union rules were in violation of the SPS Agreement was overturned by the Appellate Body—that Body held that the use of these as feed additives for growth promotion was, indeed, a "comparable situation." But the use of this was overturned because the United States could not demonstrate negative international trade effects that were a consequence of the discrimination under these "comparable situations."

[63]In the remaining dispute (Japanese fruits and nuts) a comparable situation was also available—year-to-year variations within a variety in the efficacy of fumigation techniques. Those within variety differences were comparable in magnitude with differences across varieties. However, that case was decided mainly because the fumigation requirement was more restrictive of trade than necessary and thus a violation of Article 5.6—assessment of the risks of the different possible SPS measures revealed that the varietal testing requirement achieved no appreciable reduction in risk but did substantially increase the barriers to trade.

[64]For some interpretation of the scope of "comparable," see footnote 41.

unjustifiable distortions of trade.[65] However, there is no agreed standard for how large a distortion of trade is necessary for a level or measure to qualify for the discipline of the SPS Agreement. With globalization, practically every national measure has an effect on trade, which would suggest that all SPS levels and measures could be subjected to the discipline of the SPS Agreement. However, in the EC meat hormones case, the Appellate Body appeared to decide that trade distortions must be clear and severe. It invalidated comparison between the meat hormone ban and the EC's rules that permit use of carcinogenic feed additives (carbadox and olaquindox) as "comparable situations" because the trade distortions caused by this difference in SPS protection levels were unclear and not demonstrably large (see footnote 41 and text). Furthermore, Article 5.6 of the SPS Agreement requires that members ensure that SPS measures "are not more trade-restrictive than required" (Article 5.6). Yet there is no clear standard for what is "more trade restrictive."

As the literature on national risk management shows, often risk trade-offs occur on many dimensions and are not simple. As the SPS system develops, there may be attention to trade-offs between national risks and the costs and benefits to international trade. Should a nation be urged or required to lower its SPS protection slightly, if that would allow use of much less distortionary SPS measures? Logically there is only a small step from the existing SPS Agreement—which gives countries the right to determine their "appropriate level of SPS protection"—to one that requires nations to consider international trade effects when determining SPS protection levels. Politically, it would be a huge leap.

The problem of making fine determinations in the allowable degree of trade distortion is unlikely to generate many WTO disputes for one simple reason: There must be a strong and apparent trade effect for a complaining country to justify the cost of raising and prosecuting a dispute. Vague, indirect, and secondary effects could be numerous, but they are difficult to demonstrate and are typically diffused across many actors. In contrast, identifiable and concentrated costs have a larger effect in mobilizing harmed exporters to seek change, especially if the exporters are already organized and have ready access to the agents (governments) that can prosecute the case.

Mindful of these three areas of continued uncertainty, I now turn to major conclusions that can be drawn about how the WTO system has imposed discipline on SPS measures to date.

---

[65]The Appellate Body in the EC meat hormones case underscored that the requirement of comparable levels of SPS protection in comparable situations was only in areas where the levels of SPS protection affect trade (Article 5.5). That is, all three parts of the three-part test must be satisfied. See footnote 41 and text.

No "Race Toward the Bottom" or "Downward Harmonization"

Putting discipline on SPS measures that affects trade is but one example of the interaction between the international trading system and national regulation. Freeing trade requires reducing barriers, including nontariff barriers such as national regulations, which has led many observers to fear that free trade could cause a "race toward the bottom." They fear that governments, keen to promote their exporting industries, will dismantle environmental regulations. Many have also feared that promoting the use of international standards in order to yield a level playing field will require "downward harmonization" toward the lowest common denominator. Both fears are important to examine, not only for a proper assessment of the SPS Agreement's effect on national policy but also more generally because they are central issues in the "trade and environment" debate.[66] The experience with the SPS Agreement can thus help illuminate whether and how similar fears can be addressed when other areas of national policy collide with the international goal of promoting trade by restricting nontariff barriers to trade.

The experience to date gives no support for either fear. There is no evidence of a "race to the bottom." However, it is unlikely that such a dynamic would develop in the case of SPS protection. The greatest fear with the race to the bottom is that free markets will create a strong incentive for producers to adopt less costly *processes* that make exports more competitive but have deleterious effects, such as greater pollution, in the locality of production. For SPS issues, however, the importing country retains the power to set its own level of protection, and that choice does not affect environmental quality in the exporting country. Direct competitive pressure to relax that level is largely nonexistent.

Fear of downward harmonization is potentially more valid. Indeed, the main critiques of the SPS Agreement from public interest groups have claimed that the agreement will lead to downward harmonization. Detractors further fear

---

[66]The trade and environment debate—which concerns whether trade will harm efforts to protect the environment—is not restricted to these two issues. A third major issue is whether free trade rules could undermine nations' ability to enforce international environmental agreements by use of trade sanctions. The current study is unable to illuminate that issue. The SPS Agreement encourages use of trade sanctions for the purpose of enforcing international agreements—the *Codex*, OIE, and IPPC rules are all codified in international agreements—but that is because those international agreements exist for the purpose of promoting trade. In contrast, the fear in the trade and environment debate is that a wide range of agreements that do not have trade promotion as their central goal—such as agreements to protect endangered species or reduce atmospheric pollution—will be barred from using measures that distort free trade. Many advocates of environmental protection accurately believe that threats of trade sanctions have had a large impact on forcing some countries to comply with international environmental agreements. In the future, more demanding international agreements— where incentives to cheat are higher—will require even stronger enforcement tools and thus perhaps greater use of sanctions.

the transfer of decision-making authority to international standard-setting bodies such as the *Codex Alimentarius* Commission, which they argue are undemocratic and captured by industrial interests (Silverglade, 1998; Jacobsen, 1997). Public interest groups rightly worry that their voices will not be heard when the *Codex* determines standards—with few exceptions, they have been poorly represented at *Codex* meetings, especially technical meetings where most of the detailed work of designing standards occurs (Victor, 1998). However, downward harmonization has not been much evident in practice. Legally, there is no requirement for harmonization. Rather, the SPS Agreement itself clearly states that it is for countries themselves to determine their own "appropriate level" of risk and to adopt SPS measures as necessary to meet that level. The political sensitivity of the WTO system—especially the Appellate Body—to this public fear of harmonization is apparent as it makes sure that WTO agreements are not interpreted as a guise for dismantling legitimate regulation. The decisions in all three SPS-related cases have carefully underscored the sovereign right of nations to determine their own level of SPS protection. The WTO Dispute Panel in the first case—on the EC meat hormones ban—interpreted the SPS Agreement narrowly and suggested that the inconsistency between the EC measure and the *Codex* standards was evidence that the EC measure was incompatible with the SPS Agreement. If that Panel interpretation had held, then there would indeed be reason to fear that the SPS Agreement would force strict harmonization with international standards, perhaps resulting in some downward harmonization. But the Appellate Body explicitly overturned the Panel's reasoning as too narrow; instead of strictly requiring conformity with international standards, the Appellate Body decided that countries had considerable latitude in setting SPS levels and measures that were different from international standards. The Appellate Body also set high hurdles for judging that a country's SPS levels and measures were incompatible with the SPS Agreement. The final decision remained the same—the EC ban was declared incompatible—but the reasoning was much more permissive of deviations from international standards.[67]

The three disputes might also be cited as evidence that the existence of the SPS Agreement and dispute resolution, at least in these three cases, has resulted in lower SPS protection. Indeed, in all three cases the SPS measures were struck down. However, in all three cases alternative measures offered a similarly high level of SPS protection with a much lower impediment to trade. Nonetheless, perhaps allowing any marketing of hormone-grown beef in Europe, for example, represents a reduction in SPS protection due to the SPS Agreement. However, if the science is sound—and on this there is a remarkably broad consensus among scientists—then allowing the beef on the market represents no objective reduction in SPS protection.

---

[67]As reviewed above, the body declared the EC measures illegal mainly because there was no rational relationship between the risk assessments and the level of SPS protection adopted, and the particular measures adopted were more trade restrictive than necessary.

The fact that all three of the SPS disputes to date have resulted in SPS measures being barred is not evidence that the system is instinctively opposed to strict national SPS protection. Rather, it is the product of a dispute resolution system that is both at its early stages and costly to invoke. In such a system we should expect that all or nearly all disputes raised will be settled in the complainant's favor.

What is perhaps more surprising is that more of these disputes are not settled prior to a Panel decision. Part of the explanation is that dragging out a dispute can afford an additional two to three years of protectionism. A more important explanation may be found in the EC meat hormones case. The ban on hormones is a highly politicized issue in Europe, exacerbated by the controversy over mad cow disease and genetically modified foods. A hot issue for nearly two decades, the coalition of consumer and protectionist interests are deeply entrenched. And the interests became more entrenched with time. The bans did not begin as protectionist measures but rather the responses of some EC governments to uncertain science and consumer outrage. But, over time, beef producers realized that the ban served their interests as well as did managers of the EC agriculture subsidy system, which would be even more bloated if beef production were more efficient. Politically the EC and member governments could not afford to succumb easily to international pressure.

Especially where food is concerned, these highly politicized issues will be commonplace. But a special challenge for the WTO system will be for governments to ready their societies to accept defeat and implement reform. It probably has made matters worse that the EC painted itself further into a corner by continuing to argue that European consumers required the ban to restore confidence in the market and then fanning consumer fears of the same. It is plausible that the EC has learned the lesson that this is a bad strategy because it makes the eventual fall longer and harder. Today, in the handling of import bans for genetically modified foods (which are not currently the subject of a WTO dispute, but are being debated in the *Codex Alimentarius* Commission and have been the subject of numerous risk assessments), the EC has taken a stance that is more favorable to international trade. Instead of outright bans, it favors tight regulation coupled with labeling.[68] Governments have also learned how to play the "SPS game" so as not to run afoul of trade laws. The SPS Agreement allows strict provisional measures when risks are uncertain, provided that they also make an effort to reduce uncertainties (Article 5.7). A form of the "precautionary principle," this provision allows governments to defend strict measures. The EC hormones case could not use that defense because final rules

---

[68]Whether and how labeling schemes—which in effect let consumers make import choices individually—will be ruled compatible with the WTO agreements is unclear. There are some international guidelines for labeling but there is also considerable evidence that labeling schemes can be misused as de facto trade restrictions. This issue is probably outside the scope of the SPS Agreement and, instead, is a "technical barrier to trade" and subject to the WTO "Agreement on Technical Barriers to Trade."

had been adopted before the WTO dispute was filed. Look now for governments to adapt few "final" rules when their SPS measures could be contested.

Internal Alignment of National SPS Policy Processes and Risk Levels

Thus the SPS Agreement has not required weakening of SPS measures that countries apply to protect humans, animals, and plants. But it may have a different, much larger effect on how countries manage risks. In all three of the disputes one of the critical complaints has been that import bans were arbitrary—challengers argued that the importers had required less restrictive SPS protection in other comparable situations. Although not all of those complaints were successful, the intense focus on ensuring comparable treatment has put all members of the world trading system on notice that they must be able to justify SPS regulations that were previously regarded as purely internal policy matters. Exporters have at their disposal a powerful mechanism—the SPS Agreement backed by the strong WTO dispute resolution system—with which they can attack SPS measures that are inconsistent with the rest of a nation's SPS protection scheme.

If countries are under constant pressure to justify that they adopt comparable SPS measures in comparable situations, then they are likely to give much greater attention to internal alignment of risk assessment and management policies—in other words, they are more likely to ensure that they impose comparable levels of risk management in comparable situations. They are also more likely to ensure that the particular measures they impose are based on risk assessment. The consequences of these external pressures will include much greater application of risk assessment and more transparent national SPS rules. That could be a boon for those who advocate the making of public policy according to sober assessment of risks.

(The requirement for greater use of risk assessment may make it transparent that, in the extreme, the concept of a "comparable situation" logically requires a very broad interpretation. How does one square policies that tolerate high levels of risks in one area—such as from air pollution—with *de minimus* standards for many food hazards? These are perennial questions in risk management, but the SPS Agreement may now bring them to the WTO.)

However, the impact of much greater transparency and internal alignment of risk management *on trade* is not obvious. Greater transparency should facilitate trade by making it easier for importers to identify and comply with applicable rules. Greater transparency will also make it easier for exporters to declare that they have imposed SPS measures on their exports that achieve the same level of SPS protection that the importing country achieves with (possibly different) SPS measures that it requires. If so, Article 4 of the SPS Agreement (*Equivalence*) requires the importer to let the goods flow. In democratic societies, more transparency may also make governments less likely to adopt rules that would be embarrassing and vulnerable to attack. The requirement that

SPS measures not be more trade restrictive than necessary should also facilitate trade.

But greater use of risk assessment, by itself, could have mixed effects on trade. Internal alignment of risks will eliminate grossly protective SPS measures, which should open trade.[69] But it may also result in countries tightening some SPS measures to ensure that overall national SPS policy is in alignment. One of Australia's main responses to the argument that allowing imports of potentially disease-carrying live ornamental fish was incompatible with their ban on imports of fresh and frozen salmon was to point out that it was reviewing the rules that govern imports of ornamental fish (and other potential disease carriers) (WTO, 1998b, para. 4. 190). Similarly, the EC's response to the inconsistency between allowing the use of known carcinogens (carbadox and olaquindox) while prohibiting hormones used for growth promotion was to underscore that the carcinogens were under review and might be regulated more tightly (WTO, 1998d, para. 234).

Thus, internal alignment could raise or lower SPS protection, and more analysis is needed to uncover which effects will occur. The net effect of internal alignment may greatly increase "good government" and yet have remarkably little effect on trade. More assessment should make it easier to identify SPS measures that achieve a given level of protection with fewer restrictions on trade. But more assessment may also aid the development of alliances between advocates of tight SPS protection and others who want to restrict trade, which could lead to more SPS-related trade barriers. More attention to SPS issues worldwide should lead to an international learning process focused on risk management. However, that process might result in learning about some SPS measures that are deemed "legitimate" under the SPS Agreement and are also particularly effective protectionist trade barriers—for example, quarantine measures imposed by countries that are free of particular diseases allow the importer to impose "equal treatment" on both local and foreign production while effectively barring imports. Nations may also learn that protectionist measures will not run afoul of the WTO if they also serve plausible SPS protection goals. In the EC hormones case, for example, the Appellate Body ultimately did not declare that the inconsistency between allowing the use of known carcinogens and banning growth hormones was incompatible with the SPS Agreement. The Appellate Body maintained that the inconsistency was "arbitrary or unjustifiable," but proof had not been offered to show that the difference failed the requirement of Article 2.3 of the SPS Agreement: "sanitary ...measures shall not be applied in a manner which would constitute a disguised restriction on international trade." The hormone ban had multiple effects—some legitimate

---

[69]Of course a nation could align risks so as to support a grossly protective measure. But I discount that possibility for two reasons. One is that it would require massive distortion of trade, perhaps across many sectors, which would become apparent and vulnerable to challenge both in internal political processes as well as through the WTO. The other is that even if SPS risks are aligned internally they must be based on a risk assessment (SPS Agreement, Article 5).

and others protectionist—but there were plausible reasons for the measure. For example, the Appellate Body found that "depth and extent of the anxieties experienced within the European Communities concerning ...the carcinogenicity of hormones [and] ...scandals relating to black-marketing and smuggling of prohibited veterinary drugs..." were legitimate reasons for regulatory action (WTO, 1998d, para. 245). Moreover, the Appellate Body also judged as reasonable the fact that the EC had banned hormones across the community—although only a few EC members had taken such action on their own—because the goal of a barrier-free internal market required a common measure across the EC. Whereas the Panel had focused on the protectionist reasons for the inconsistency, the Appellate Body underscored the legitimate ones. By this reasoning, the Appellate Body thus set a high standard for complainants to meet when they attempt to prove that a measure has protectionist aims. The lesson learned by protectionists is certainly to shroud a protectionist measure with legitimate food safety concerns. Expect bootleggers to seek out (or invent) Baptists to help press their interests.

More generally, increased attention to evaluating risks could result in a greater number and diversity of SPS measures. As societies have become more aware of risks and better able to afford risk management, they have demanded more stringent social regulation. Within this context, international rules that force countries to look more closely at their SPS policies are likely to yield more SPS measures by accelerating the tendency for countries to impose SPS measures. And the SPS measures that countries do adopt are more likely to be tuned to local conditions and interests if they are explicitly based on risk assessment. It is thus plausible—perhaps even likely—that the result of greater attention to SPS measures will be *greater diversity* in SPS levels and measures, not harmonization.

More systematic analysis is needed to determine the trade effects of the SPS Agreement's effort to harmonize the process of SPS protection—in particular, the requirement that members employ risk assessment. It should not be assumed that in all conditions a binding requirement to base SPS measures on risk assessments will lead to more trade or even to more harmonization.

International Standards: Little Impact

Although the central purpose of the SPS Agreement is to promote harmonization of SPS measures, the three cases suggest that international standards have not had much impact. That finding is surely biased by the cases examined here—dispute panels are likely to hear only those cases for which national SPS measures do not conform with international standards since cases where there is conformity are explicitly in compliance with the SPS Agreement and thus yield no viable dispute. Thus, perhaps international standards are having a large unseen effect; systematic research on that possible effect is needed.

In each of the three WTO cases, international standards were referenced in the resolution of the disputes. But none of the outcomes from the disputes actually required the existence of an international standard. The EC hormones case made most extensive use of international standards, but that was because the *Codex* system—in particular JECFA (which is formally external to *Codex*)—had given extensive attention to the hormones under review. Even so, the dispute panels did not rely exclusively on the JECFA reviews. Rather, the Panel (advised by experts it had retained) looked at the entire scientific literature, which included several non-JECFA reviews of hormone risks. The JECFA reviews were helpful and set a clear benchmark for quality scientific assessment, but the other scientific reviews came to the same conclusions. Moreover, by overturning the narrow interpretation of the SPS Agreement as requiring *conformity* with international standards, the Appellate Body underscored that international standards were at best a starting point for countries that wanted to deviate from them. Indeed, the existence of international standards was irrelevant for the main line of legal reasoning that decided the EC meat hormones case—the failure for the EC to establish a "rational relationship" between risk assessment and the measures it imposed. The lack of any international standard for one of the six hormones (MGA) did not excuse the EC from the obligation to base even its ban of that hormone on a risk assessment (see footnote 39 and text above).

The minimal influence of international standards is even more evident in the Australian salmon and the Japanese fruits and nuts cases. In those cases the OIE and IPPC, respectively, had few, if any, standards that were directly applicable to the issues in the disputes. Only a few of the fish diseases were on the lists of diseases in the OIE's *International Aquatic Animal Health Code*, and thus for only those did OIE offer specific guidelines for imposing trade restrictions. For the other diseases, OIE was largely silent. Both OIE and the IPPC promulgated general standards for risk assessment that could be applicable in those cases where more specific international methods and standards did not exist, but those guidelines were so broad as to be essentially irrelevant to the resolution of these two cases.

This suggestion—that international standards have had much less importance than expected—may hold in the future as well, but for an additional reason. Greater internal alignment of risk management procedures (see above) need not result in alignment according to international standards. Indeed, because the SPS Agreement allows liberal deviations from those standards, more systematic national risk assessment and greater public debate over acceptable risks are likely to result in many more deviations. The history of the *Codex Alimentarius* acceptances (Table 6-1) lends support to this argument. Developing countries lodged more "full acceptances" of *Codex* commodity standards because they did not have many SPS measures already in place. But industrialized countries—especially those with the most advanced SPS protection systems—employed principally "acceptances with specific deviations." This suggests that international standards are a fluid that can fill gaps (when countries let the fluid flow), rather than a solid block that crushes deviation. Moreover, this history suggests that the natural progression of risk management may be toward diversity, not harmonization.

In a decade from now it may be clear whether the assessment suggested here has proved robust—that the SPS Agreement is having a large effect on the process of national risk management without much increasing the degree of harmonization with international standards. That experience may help build consensus around the view that the solution to nontariff barriers to trade is *discipline*, not *harmonization*. Whether discipline alone will be enough to achieve adequately free trade remains to be seen. But it could be adequate to blunt opposition to trade liberalization that has, in part, been based on the fear that free trade requires harmonization.

International Standards: From Artificial Consensus to Conflict

All three of the international standard-setting bodies examined in this paper were created for different purposes and then grafted to the WTO system. The discussion here focuses on the *Codex Alimentarius* Commission—the most important of the three. *Codex* was originally created to provide a forum that would facilitate coordination and create standards that countries would implement voluntarily. The SPS Agreement has changed that dramatically by making the standards legally binding and enforceable.

The change in status is bound to lead to greater conflict. The consensus and ease of operation that prevailed in the past was artificial—it existed because the standard-setting bodies were not particularly relevant and entirely optional. Today the stakeholders believe that the standards are much more relevant than before. They might be wrong (see above), but it is perceptions that matter. The newfound importance of these international bodies is requiring much greater attention to decision-making procedures and also exposing standards organizations to greater conflict.

The requirement in the SPS Agreement that SPS measures be based on risk assessment unless they are based on international standards will underscore the need for international standards themselves to be based on risk assessment. Yet, to date, none of these organizations has applied a systematic policy for determining acceptable levels of risk. The *Codex* is in the midst of a system-wide reassessment of risk-related concepts and procedures. But progress has been very slow. Before the conclusion of the SPS Agreement, the *Codex Alimentarius* Commission contained no principles or definitions related to the application of risk assessment and risk management (*Codex Alimentarius* Commission, 1993). Today it has several general statements on the role of science and risk assessment, and efforts are under way to expand treatment of risk in setting *Codex* standards.[70] However, the *Codex* risk principles and

---

[70]For the first amendment to the *Codex Procedural Manual* that adds these statements see *Codex Alimentarius* Commission, (1997), paras. 26–31 and pp. 90–91).

definitions are only broad statements that appear unlikely to have much practical impact on the work of *Codex* or on how cases under the SPS Agreement might be handled. It is essential that *Codex* continue development in this area, but it is also likely that debates over the risk principles will underscore that in the past the *Codex* standards were easily adopted because there had not been a rigorous attempt to align the risk levels across all standards. Trying to do so would have revealed that gaining agreement on a single world level of acceptable risk—which, in turn, could determine *Codex* standards—is difficult or impossible.[71]

To date, the *Codex* standards have largely reflected risk management procedures in the advanced industrial countries. They are developed with extensive input from industry, mainly (but not exclusively) in the advanced industrialized countries. The industry's interest is to ensure that international standards are consistent with national practices. Thus they seek international standards that mirror those at home. The result is that the agenda and standards in the *Codex* are determined heavily by the SPS policies in the advanced industrialized countries. Similarly, the large safety margins and the desire to set standards at the "no effect" level reflect the goal of the advanced industrialized countries, which is to set food safety risks as close to zero as is practical. It remains unclear whether that view of the proper risk level is shared among all *Codex* members. Until the application of *Codex* standards through the SPS Agreement, few *Codex* members had paid close attention to the exact safety levels that the *Codex* system assured.

Rising conflict in standard-setting bodies should not be lamented. It is the by-product of a shift from a voluntary (often ineffective) system of standards to a scheme that may have more binding impact. That shift has made some players less willing to sacrifice their interests for the sake of agreement. Previously, compromise was less painful because the *Codex Alimentarius* procedures, especially the provisions for acceptance, were rife with opportunity to opt out of inconvenient commitments. The impression that international standards are now more relevant has also entrained new actors—such as consumer protection organizations—into the process, and with new voices and interests it has proved more difficult to reach consensus or even majority agreement.

But it is worthwhile to ask whether the strategy adopted has been the most effective. In the case of the GATT, it was the GATT members themselves that made the shift from the weaker GATT 1947 framework to the integrated WTO system. They changed not only the legal framework but also the stringency of commitments and the mechanism for enforcement. In the case of the *Codex,* however, the change in de facto legal status has arrived on its doorstep from the

---

For current efforts to improve definitions see *Codex* Committee on General Principles (1998).

[71]It might be easier to adopt regional levels of acceptable risk because nations within a geographical region might have common views of risk. However, regional *Codex* decision-making processes—which were allowed before the *Codex* system was incorporated into the WTO—are no longer allowed. Nor would regional standards be consistent with the WTO's principles of universal access to decision-making procedures.

outside; internal *Codex* procedures and mechanisms were not reformed at the same time. With *Codex* standards now much more relevant, is the huge committee system with majority voting the most efficient way to adopt standards? If, in the future, other standards organizations are brought within a binding framework, is there anything that can be done to prepare the way?[72]

Finally, although conflict can be productive it does risk greatly slowing the work of international standard-setting bodies because nations (or at least a majority of them in each case) are less willing to adopt standards that could be relevant. The failure of the *Codex Alimentarius* Commission to adopt a standard for bovine somatotropin—for which scientific evidence of safety is comparable with that of growth hormones—suggests that, on politically charged topics, the standard-setting bodies may become bogged in deadlock. The result will be a lack of standards (or broad and meaningless guidelines that are equally useless). That deadlock may not matter for the disciplines of the SPS Agreement because the requirement to base national SPS levels and measures on risk assessment (Article 5) remains in place, even when international standards do not exist.

However, on balance, the deadlock on standards probably hurts the free trade agenda. The one area where international standards have been consistently influential is when filling gaps—in areas of food law and in nations not already covered by standards. As markets open, the number of gaps—especially in countries where administrative capacity is low—will grow, at least in the short term until countries catch up with the process of national risk assessment and management. International standards could thus play an especially important role in opening trade to new markets, new products, and new methods of SPS protection. Examples currently on the agenda of the WTO include genetically modified organisms, labeling, and a scheme for more consistent implementation of SPS measures known as Hazard Analysis and Critical Control Points. If nations are gridlocked in *Codex* because they fear binding application in the WTO, they will not have adequate international standards to guide their efforts to address new SPS threats and new opportunities for improved SPS protection.

Expert Advice

The experience with SPS discipline offers several encouraging stories for the perennial problem of incorporating expert advice into the processes of risk assessment and risk management. This problem has often been poorly handled within countries, especially when the need for expert advice intersects with

---

[72]There are some efforts at internal alignment already under way in the *Codex* system. Examples include the development of principles for risk assessment (mentioned above) and the development—now far advanced—of new simplified commodity standards that are intended to focus commodity standards on the food safety-related aspects of commodities. (Prior to this effort, *Codex* commodity standards were not developed according to a uniform procedure, and many standards addressed cosmetic and other attributes of foods that had little to do with safety.)

adversarial legal proceedings. Often the result is a duel between expert advocates that buries and confuses "the truth."

The experience reviewed in this paper confirms that independent expert panels, in contrast with dueling experts, can be a highly effective device for synthesizing complex technical information. The *Codex* committees that debate the proper level for MRLs would have found it very difficult to agree on standards without the advice of the JECFA and JMPR expert committees. Experts have also played a vital role in the resolution of the three WTO disputes. In each dispute, the WTO dispute panels enlisted experts to answer technical questions.[73] Although it is difficult to assess what would have happened if the dispute panels had not had access to their own experts, it is clear that the experts' assessments formed a prominent part of the record and decision-making process in each case. Approximately one-fourth of each Panel report consists of the transcript from the consultation with expert panelists. Each of the three cases depended critically on expert interpretations of scientific evidence as well as evaluations of the adequacy of risk assessments. The central issue for the Panel in the meat hormones case was to determine the merit of the EC's claim that there were consumer risks associated with using hormones for growth promotion. In the salmon case, two critical judgments relied heavily on technical information—whether headless and eviscerated fish posed disease risks, and whether other imports (e.g., frozen bait herring) posed comparable risks. In the Japanese fruits and nuts case, the critical issue was the highly technical conflict over the efficacy of fumigation techniques.

The SPS Agreement may also change the type of information that is demanded of experts and sought through the risk assessment process. All three of the WTO panel cases made extensive use of risk assessments. But each case required not only the "normal" elements of risk assessment—an evaluation of the risks to humans, animals, or plants in the importing territory—but also an assessment of how different measures to manage those risks would affect trade. As a growing number of national SPS measures come under scrutiny for their consistency with the SPS Agreement, this "trade impact assessment" aspect of risk assessment will probably become commonplace.

SUMMARY

In this paper I have reviewed the provisions of the 1994 SPS Agreement and the first three WTO disputes related to the application of the SPS Agreement. I have argued that large areas of interpretation remain open.

---

[73]The SPS Agreement empowers dispute panels to seek advice from experts chosen by the Panel, establish an advisory technical experts group, and/or engage in other expert consultations (Article 11.1). The WTO *Understanding on Rules and Procedures Governing the Settlement of Disputes* allows panels to request an advisory report from experts on scientific or technical matters (Article 13.2 and Appendix 4). None of the three cases here have employed this provision; rather, in each case, the more interactive format of expert consultation specific to the SPS Agreement was used.

However, the disputes have allowed some interpretation of all the major obligations of the SPS Agreement. In particular,

• The SPS Agreement urges countries to apply international standards, but it explicitly allows countries to deviate from those standards if they can justify the deviation. However, SPS policies that deviate from international standards must be based on risk assessment. In the Australian salmon case, the Appellate Body clarified the "three prongs" of a satisfactory risk assessment and demonstrated why the risk assessment in the Australian salmon ban was not adequate.

• Adequate risk assessments did exist in the two other cases (EC meat hormones and the Japanese fruits); however, in both of these cases the Panels and the Appellate Body deemed the SPS measures illegitimate because there was no "rational relationship" between the risk assessment and the measures applied.

• The SPS Agreement also requires that countries apply comparable SPS policies in comparable situations. In the EC meat hormones case the Appellate Body created a three-part test that clarified this obligation. Applying this test, the Appellate Body found that the ban was "arbitrary or unjustifiable," but it did not declare the ban invalid because there was insufficient evidence that the ban also constituted a disguised restriction on international trade (SPS Agreement, Article 5.5). In the Australian salmon case the evidence for discrimination was stronger, and Australia's ban was declared invalid.[74]

• The SPS Agreement also requires that countries apply the least trade-restrictive measures. In two cases (Australian salmon and Japanese fruits and nuts), WTO panels ruled that other measures were available that would achieve the same level of SPS protection and were substantially less restrictive of trade. The basic logic of those findings has been upheld, although the Appellate Body overturned each on other grounds.[75]

• The SPS Agreement requires that countries make their SPS policies transparent so as to facilitate compliance and trade (Article 5.8 and especially Article 7). In the Japanese fruits and nuts case, the Panel found that the transparency provisions had been violated because importers found it difficult to understand the de facto rules that governed imports into Japan, and Japan had not published the relevant material in a way that was accessible to outsiders.

---

[74]The interpretations developed through these cases are doubly important. Not only do they affect the handling of particular cases but they also may influence (or replace) guidelines that the SPS Committee is supposed to develop in order to identify which situations are comparable. That effort is under way but has not made much progress.

[75]In particular, in the Australian salmon case insufficient information was available to determine Australia's level of SPS protection because Australia had not conducted a risk assessment or evaluated alternative SPS measures. In the Japanese fruits and nuts case, the panel finding was overturned on procedural grounds—the Panel and experts had identified the alternative, less trade-restrictive measure, whereas it was the obligation of the United States to do that.

Thus, nearly all the major obligations of the SPS Agreement have been addressed, at least partially, in disputes: the urging of nations to apply international standards, the requirement to base SPS measures on risk assessment, the requirement to apply comparable policies in comparable situations, the requirement to use the least trade-restrictive measures available, and the requirement that nations make their SPS policies transparent. The only major requirement that has not been addressed in a dispute concerns "equivalence" (Article 4), which obliges each member to accept the SPS measures of other members as equivalent if the exporter can demonstrate that its SPS measures achieve the level of SPS protection required by the importer. This could prove to be the agreement's largest effect on trade, but working out what is meant by "equivalence" will be complicated.

Although many areas of interpretation remain gray, the legal text of the SPS Agreement and the cases to date have underscored that nations have wide latitude in setting their SPS protection levels and measures. Thus, far from imposing a strict harmonization between national and international standards— which was the main fear of the agreement's detractors—the agreement actually allows diversity to flourish. The agreement is likely to result in increased use of risk assessment, especially in nations where risk assessment has been used only rarely or never, and to promote debate within nations about proper SPS risk management. More informed and extensive debate will likely lead to even greater diversity in SPS levels and measures.

The evidence suggests that harmonization of SPS *levels* and *measures* is not under way. However, harmonization of national SPS *procedures*, such as the requirement for risk assessment, is evident. Procedural harmonization without the strict requirement for harmonization of levels and measures may help to mute the backlash against globalization that, in part, is animated by the fear that national sovereignty is being lost to undemocratic international standard-setting bodies. That procedural harmonization is largely the result of how the Appellate Body—a standing body that serves as a steward of the WTO system and is politically more astute than the panels, which are convened for a particular case and then disbanded—has interpreted the SPS Agreement. In particular, the Appellate Body overturned the legal reasoning of the Dispute Panel in the hormones case, which had maintained that national SPS measures must conform with international standards. The Appellate Body's interpretation probably gives nations more latitude than the creators of the SPS Agreement had originally intended. Advocates of international rule of law probably lament that outcome, but it seems to have been politically wise.

Even procedural harmonization has been difficult to implement. At this writing, the United States and Europe are in the midst of a trade war over meat hormones—a case that Europe lost not because it was forced to don the straitjacket of international standards but because it could not demonstrate that there was any "rational relationship" between risk assessments and the standards it had adopted. To stay on an even keel, the WTO may need additional devices to serve as a safety valve when international rules come into direct conflict with entrenched national interests. With the WTO agreements in 1994, overnight members of the world trading system shifted from a system in which there was essentially complete freedom to adopt numerous national SPS measures that affected trade to one where freedom is constrained by the SPS Agreement. The transformation has not been easy and is not complete; if not managed with political sensitivity to the domestic backlash it could readily derail.

REFERENCES

Boardman, R. 1986. Residues of pesticides: the Codex system. Chapter 4 in Pesticides in World Agriculture: The Politics of International Regulation. New York: St. Martin's Press.

Charnovitz, S. 1997. The World Trade Organization, meat hormones, and food safety. International Trade Reporter 14(41):1781–1787.

Codex Alimentarius Commission. 1993. Procedural Manual. Rome: Joint FAO and WHO Foods Standards Program. Rome: FAO.

Codex Alimentarius Commission. 1997. Report of the 22nd Session, Geneva, 23–28 June. Rome: FAO.

Codex Committee on General Principles. 1998. Risk Analysis: Working Principles for Risk Analysis. CS/GP 98/4, Thirteenth Session, Paris, France, 11–15 May 1998. Rome: FAO.

Hudec, R.E. 1993. Enforcing International Trade Law: The Evolution of the Modern GATT Legal System. Salem, Mass.: Butterworth Legal Publishers.

Jacobson, M.F. 1997. Consideration of *Codex Alimentarius* standards, advance notice of proposed rulemaking. Comments of the Center for Science in the Public Interest. Docket 97N-0218. U.S. Department of Health and Human Services and Food and Drug Administration, Washington, D.C.

Kay, D.A. 1976. The International Regulation of Pesticide Residues in Food. Washington, D.C.: American Society of International Law.

Leive, D.M. 1976. International Regulatory Regimes: Case Studies in Health, Meteorology and Food, 2 vols. Lexington, Mass.: Lexington Books for the American Society of International Law.

Maddox, J. 1995. Contention over growth promoters. Nature 378:553.

OIE (Office International des Epizooties). 1997. International Aquatic Animal Health Code, 2nd ed. Paris: OIE.

OIE (Office International des Epizooties). 1998. International Animal Health Code, 7th ed. Paris: OIE.

Roberts, D. 1998. Preliminary assessment of the effects of the WTO Agreement on sanitary and phytosanitary trade regulations. Journal of International Economic Law 1(3):377–405.

Roberts, D., and K. DeRemer. 1997. Overview of Foreign Technical Barriers to U.S. Agricultural Exports. ERS Staff Paper No. 8705. Economic Research Service, Commercial Agriculture Division. Washington, D.C.: U.S. Department of Agriculture.

Silverglade, B.A. 1998. The impact of international trade agreements on U.S. food safety and labeling standards. Food and Drug Law Journal 53:537–541.

Tutwiler, A. 1991. Food Safety, the Environment and Agriculture Trade: The Links. Discussion Papers, Series No. 7. Washington, D.C.: International Policy Council on Agriculture, Food Trade.

Victor, D.G. 1998. The operation and effectiveness of the Codex Alimentarius Commission. In Effective Multilateral Regulation of Industrial Activity: Institutions for Policing and Adjusting Binding and Nonbinding Legal Commitments. Ph.D. thesis, Department of Political Science, Massachusetts Institute of Technology, Cambridge.

Vogel, D. 1995. Trading Up: Consumer and Environmental Regulation in a Global Economy. Cambridge, Mass.: Harvard University Press.

WTO (World Trade Organization). 1997a. EC Measures Concerning Meat and Meat Products (Hormones), Complaint by Canada. Report of the Panel, WT/DS48/R/CAN. Geneva: WTO.

WTO (World Trade Organization). 1997b. EC Measures Concerning Meat and Meat Products (Hormones), Complaint by the United States. Report of the Panel, WT/DS26/R/USA. Geneva: WTO.

WTO (World Trade Organization). 1998a. Australia—Measures Affecting Importation of Salmon. Report of the Appellate Body, WT/DS18/AB/R. Geneva: WTO.

WTO (World Trade Organization). 1998b. Australia—Measures Affecting Importation of Salmon. Report of the Panel, WT/DS18/R. Geneva: WTO.

WTO (World Trade Organization). 1998c. EC Measures Concerning Meat and Meat Products (Hormones). Arbitration under Article 21.3(c) of the Understanding on Rules and Procedures Governing the Settlement of Disputes, WT/DS26/15, WT/DS48/13. Geneva: WTO.

WTO (World Trade Organization). 1998d. EC Measures Concerning Meat and Meat Products (Hormones). Report of the Appellate Body, WT/DS26/AB/R, WT/DS48/AB/R. Geneva: WTO.

WTO (World Trade Organization). 1998e. Japan—Measures Affecting Agriculture Products. Report of the Panel, WT/DS76/R. Geneva: WTO.

WTO (World Trade Organization). 1999. Japan—Measures Affecting Agriculture Products. Report of the Appellate Body, WT/DS76/AB/R. Geneva: WTO.

7

Accounting for Consumers' Preferences in International Trade Rules[1]

JEAN-CHRISTOPHE BUREAU

Institut National Agronomique Paris–Grignon and Institut National de la Recherche Agronomique (INRA), Grignon, France

STEPHAN MARETTE

Institut National de la Recherche Agronomique, Grignon, and Thema Université Nanterre, France

SANITARY AND TECHNICAL BARRIERS

National Regulations as Trade Barriers

The Uruguay Round has led to a substantial reduction in tariff protection. As traditional trade barriers tend to come down, nontariff trade barriers are becoming a more important issue in the agriculture and food sector. This includes sanitary regulations and, more generally, a larger set of technical rules embedded in national regulations. A few years ago, little was known on the trade effect of domestic regulations, but this issue is now becoming more and more documented. The listing of barriers to U.S. exports by the Economic Research Service of the U.S. Department of Agriculture (USDA) is one of the most detailed works in this area (Roberts and DeRemer, 1997; Thornsbury et al.,

---

[1]Although the scope and content differ a good deal, this paper uses sections of a report prepared for the Organization of Economic Cooperation and Development by J. C. Bureau, E. Gozlan and S. Marette. This document was, subsequently, revised by W. Jones and published as a consultant's report, *Food Quality and Safety, Trade Considerations* (1999), under the responsibility of the Secretary General.

1997). To our knowledge, it is so far the only large-scale attempt to quantify the effect of foreign sanitary and technical regulations on trade. Other agencies, such as the U.S. International Trade Commission and the Office of the U.S. Trade Representative, have also investigated the impact of many national regulations on exports. Both the European Union (EU) and the Canadian government publish an annual report on U.S. trade barriers. The European Commission has also set up a database on market access for a large number of countries. This database describes many foreign regulations that European exporters consider as unnecessary barriers to trade.

These efforts show that domestic regulations impede imports in almost all countries. Regulatory barriers in the EU are often pointed out by U.S. agencies. The EU ban on hormone-treated meat is one of the most quoted examples. In the European Commission's market access database, the pages relative to Japan are particularly impressive. Many regulations, from the list of authorized additives to the technical requirements on meat products and the conditions of fumigation of flowers and vegetables at the Narita airport, are considered excessive. Even Australia, a country known for low tariffs, has technical standards that often preclude imports.[2] The U.S. conditions of sanitary inspection, with long and somewhat random delays, open lists for insects which make import authorizations difficult and unpredictable, and complex quarantine rules are also accused of making it unnecessarily difficult to export food products to the United States.[3]

International rules have been strengthened to address these problems. The Uruguay Round provides a framework for solving disputes through the World Trade Organization's (WTO's) Dispute Settlement Body; it tackles the problem of nontariff trade barriers through the Sanitary and Phytosanitary (SPS) Agreement and a strengthened Technical Barriers to Trade (TBT) Agreement; and it gives greater importance to international bodies, especially *Codex*

---

[2]According to the EU Commission, "the fervor with which sanitary and phytosanitary rules are applied in Australia suggests that the system operates as a trade barrier." For example, canned tomatoes that have failed a case pressure test are banned; and it is nearly impossible for cereals to enter Australia even at times of severe drought (when domestic prices are very high) due to disease fears. Import permits are required for over 150 agricultural products, without which they are prohibited, and the significant financial costs of product control and testing, as well as the slowness of the monitoring process, serve to deter trade of chocolate, canned meat, olives, wine, herbs, poultry, or pork meat. Quarantine regulations de facto prohibit the importation of a whole range of meat, dairy, and other products.

[3]The Organization for Economic Cooperation and Development (OECD) noticed that there are 11 U.S. agencies involved in import regulations, many of them with different methods of assessments, imposing an unnecessary administrative burden on would-be exporters to the United States (OECD, 1997a). For example, imported foods are treated differently depending on whether they are regulated by the Food and Drug Administration or the USDA (the USDA inspects meat and poultry products), and as a result, different processed products exported by the same firm are sometimes treated differently.

Alimentarius, an international code of standards for human health protection under the auspices of the Food and Agriculture Organization (FAO) and the World Health Organization (WHO). The SPS Agreement (Appendix A) recognizes the right of governments to restrict trade in order to protect human, animal, or plant health, but such measures must be transparent, consistent, and based on international standards or scientific risk assessment. There must be equal treatment for all nations and between imports and domestic products. The SPS Agreement covers health risks (food safety) arising from additives, contaminants, toxins, and pathogens contained in food products. The TBT is much broader, covering all technical regulations, voluntary standards, conformity assessment procedures, and any other measures not covered by the SPS Agreement. It seeks to ensure that national measures are transparent, have a legitimate purpose, and minimize restrictions on trade. Compliance with relevant international standards is encouraged. Recently, at the Singapore conference, the WTO has raised the issue of simplifying the procedures for imports, which are often complex and act as trade barriers. Finally, in addition to the measures taken at the WTO level, the regulatory reform proposed by the Organization for Economic Cooperation and Development (OECD) also aims at limiting the negative trade effects of national regulations (OECD, 1997a).

International Effects of National Regulations

There is no doubt that many of the regulatory barriers mentioned above, in particular SPS barriers, have been erected to protect local producers. Anecdotal evidence shows that special interest groups have often persuaded public authorities to back their case and erect barriers to protect vested interests, and that governments have sometimes "compensated" for the decrease in tariffs by stricter SPS regulations to prevent a surge in imports. The shorter shelf life for food products in Korea was a famous example, but many other cases have been reported (see Hillman, 1997, and, more generally, the different papers in Orden and Roberts, 1997).

However, governments have also often set up regulations to address consumers' concerns. Such regulations often have a negative impact on trade, although this was not their primary purpose. This is the case, for example, when various options exist for ensuring a given level of consumer protection. To ensure that a product is safe, a government may consider banning certain techniques or laying down maximum tolerance levels for residual pathogens. If one country's standards are based on the first option and another country's are based on the second, exports come up against technical barriers and additional control costs.

Differing incomes and tastes may lead to differing regulations. Developing countries cannot allow themselves the same standards as developed countries, with the result that their firms come up against regulations that constitute a de facto barrier to exports. Even in developed countries, in economic terms, it is possible to determine an optimum standard for each country, reflecting in particular a trade-off between cost and demand for food safety (Antle, 1995;

Viscusi et al., 1995). It depends on the distribution of consumers' willingness to pay, and there is no reason why such an optimum standard should be the same in all countries. But different standards, albeit entirely justified in economic terms, can hamper international trade.

More generally, regulations that affect trade may come from genuine technical, geographical, cultural and sometimes religious differences. The concept of product quality is multidimensional and is not limited to product safety (Hooker and Caswell, 1996). The perception of which attributes are essential when defining quality differs greatly among countries. Differing tastes, incomes, and willingness to pay for a particular attribute are reflected in dissimilar regulations. In many countries, there is a public debate over regulation of the food industry. This includes safety of food, and how it is produced (i.e., social conditions, animal welfare, the use of genetically modified organisms, hormones and growth promoters, cultural preferences, resource sustainability, and protection of the environment). New production and processing methods driven by technology have added to consumer unease. The resulting national regulations can pose problems for exporters. The complexity of the issues make the right policy response difficult, especially in the absence of convincing evidence of health risks, but when consumer concerns look nevertheless genuine. In the following sections we present a few examples of such controversial issues.

TECHNICAL AND CULTURAL DIFFERENCES AND DOMESTIC REGULATIONS

Disagreement on Quality Attributes

There is considerable disagreement on quality attributes, such as the nutritional content, taste, production methods, and authenticity of products, that are relevant and on the extent to which they may legitimately be the subject of regulation. Some countries consider that the soil, climate, and traditional know-how that exist in a region have a decisive influence on product quality, but others do not. There is considerable disagreement on the meaning of "authenticity," which has inspired the 1992 EU legislation on food quality labeling (the term authenticity is used to translate the concept of "typicité," which is the basis of all French and Italian quality labeling systems, meaning that a product must be "typical"—for example, representative of a particular area, in addition to being produced with premium raw materials and, often, traditional techniques). Definitions based on taste or traditional know-how receive little support at an international level. These notions of product quality are ill-matched to the more restrictive approach adopted internationally.[4] The

---

[4]Chen (1996) highlights the incompatibility of European quality marks, which emphasize authenticity, with U.S. legislation and the difficulty of achieving international recognition for this type of mark.

stance of the SPS Agreement is to take into consideration only a single quality attribute, namely sanitary quality. Labels of the International Organization for Standardization, which could become de facto standards regulating international trade, do not include all the quality dimensions of European regulations, which are based to a considerable extent on a product's organoleptic qualities (taste) and authenticity.

Different Conceptions of Risk

It is seldom possible or economically feasible to achieve zero risk with respect to food safety. The SPS Agreement explicitly requires risk assessment to be carried out if a country adopts different standards from those of the *Codex Alimentarius* (Article 5.1). However, there is no agreement on what constitutes justifiable risk or "acceptable risk" as mentioned in the SPS Agreement (Annex A.5). Nor is there any agreement on the importance to be given to risk analysis, or on what is meant by the term "risk," or on methodology. Officially, risk analysis is a three-stage process. The first stage, risk assessment, consists in identifying hazards, in particular their forms, thresholds, and probabilities. The second phase is risk management, and the third phase is communication concerning the risk. Approaches may differ widely from one country to another, especially concerning the importance to be placed on risk management (Mazurek, 1996). Some countries prefer to emphasize risk elimination (e.g., sterilization of mineral waters, a ban on cheese made from unpasteurized milk, etc.). Others emphasize the possibility of risk control (in the above-mentioned examples by bottling at source, Hazard Analysis at Critical Control Points [HACCP] controls, etc.), which is sometimes less costly and alters the product less, and point to the inconsistency of seeking to achieve zero risk in one area while tolerating high risk in others.[5]

The diverging conceptions of risk management are particularly obvious in many debates within *Codex* committees, such as the one on food hygiene or the one on dairy products. The case of cheese made from unpasteurized milk provides an illustration of the fundamental differences that exist with regard to food safety thresholds between the EU and the United States (note that this is also the case within the EU itself). Cheese made from unpasteurized milk is more likely to contain pathogenic bacteria (*Campylobacter, Salmonella, Listeria*) than cheese made from pasteurized milk. Raw milk cheeses are, however, widespread in countries like France, Switzerland, and Italy. In France, consumers clearly find that the hazard is minimal compared to other types of risk, including the risk of infection of pasteurized cheese when improperly

---

[5]The SPS Agreement states that countries should have the objective of "consistency." If, on the basis of a risk assessment, there is a one in a million chance of a certain product causing a certain level of damage, the product should not be subject to greater restrictions than other products presenting a similar level of risk. The level of risk may be acceptable or not, the objective is that the acceptable risk should not be different according to the product concerned (see Doussin, 1995).

stored, and that there is basically no danger.[6] Risk management (control of dairy processing plants) and risk communication (warning elderly people and those with a weak immune system, systematic warning of pregnant women by doctors about possible abortions) are seen as being preferable to mandatory pasteurization. That is, risks are given media coverage and people clearly accept them. Any attempt to restrict the sales of even the most potentially dangerous soft cheeses is considered as completely unacceptable, and this issue clearly becomes a quasi-religious one every time it is raised at the EU level. French, Italian, and Swiss consumers point out the inconsistency of banning raw milk cheese and not, say, raw oysters or hamburgers in other countries. However, it is clear that consumers in other countries are not willing to accept the level of risk associated with raw milk cheese, possibly because they are less sensitive to quality attributes such as "taste" and "authenticity."

Technical Regulations and Local Conditions

National regulations on authorized pesticide residues, for example, differ widely. However, the fact that it is difficult to measure the risks in this area makes any attempt to define standards highly controversial (Mazurek, 1996). Even national regulations applied evenhandedly to domestic and imported products can have an effect on trade, especially if the chemical substances are not used in the country concerned. This is the case with procymidone, for example, a fungicide that is the subject of controversy in the wine-making industry. As the fungus against which procymidone is effective does not pose a problem in California vineyards, mainly for climatic reasons, there is no reason to use procymidone there. But low tolerance levels for residues would indirectly limit imports of wine from other countries, which need to use the product because of their climate. This issue, and more generally the "Delaney clause" in the U.S. legislation, was a bone of contention between the EU and the United States. This highlights the possible trade effects of national regulations, even when they are applied to imported products in a nondiscriminatory way.

Some techniques that are used to control bacteriological risk are more adapted to certain countries than others. Bottling mineral water at source, for example, may be more expensive than bulk transportation and sterilization in some cases. HACCP techniques require sophisticated technology and qualified labor at all stages of the production, transformation, and marketing chain, which may be difficult to find in all countries. Irradiation techniques require a lot of capital, and overall it is very costly to transport products with low unit value to the adequate plant, especially in countries where production is scattered over

---

[6]Acording to the Ministère de la Santé (i.e., the health department), milk products were responsible for 5 percent of alimentary toxicoinfections, that is, 5 percent of 0.00016 percent of the meals served in 1995 (only a share of them being raw milk products); unpasteurized cheese was, however, clearly involved in the death of one person in 1997 (to our knowledge, none in 1996 and 1998).

very large areas. In many cases, different techniques (e.g., controlling processes or sterilizing) can give equivalent results at the end product, but not all of them are adapted to the domestic conditions. Again, if one country picks an option and a second country picks another one, this will de facto result in technical barriers for exporters.

The role that should be left to private operators in devising workable standards is a source of disagreements between countries. Producers want to be given greater freedom in the way they produce high-quality food and point to the costs that highly specific regulations impose on the production process. They find it hard to understand why consumers and public authorities interfere so much in the definition of standards which in other (nonfood) sectors is more commonly left up to industry. Consumers do not see things in the same light and criticize what they regard as industry's over-representation on the scientific committees of standards bodies, such as the Joint FAO/WHO Expert Committee on Food Additives, the Joint FAO/WHO Meeting on Pesticide Residues, and the *Codex* committees. The "technological justifications" that allow manufacturers to use nonclassified additives are another source of consumer concern. The same goes for the "codes of good practice" which manufacturers may invoke when introducing additives. This highlights the difficulties of finding the right mix between highly detailed and restrictive regulations and consumers' concern at the latitude accorded to manufacturers. It affects trade issues because the "right mix" is not seen as being the same in all countries, an issue raised in numerous disputes on the list of permitted food additives (Vogel, 1995). For example, European consumers argue that substances used by manufacturers to preserve the color of perishable foods (hydroxyquinone, polyphosphates) do not improve the products and may even be misleading as to their freshness. U.S. consumers seem less worried (although they seem to have more concerns on other issues, for example, on the use of aspartame).

Legal Differences

Domestic regulations are defined in relation to the legal system prevailing in each country. This framework differs a lot across countries. Punitive damages in product liability action are very different in the United States and in European countries. In the United States, ex post liability clearly plays an important role in deterring firms from marketing unsafe products. Because of the potential outcome of tort law, firms often set up standards that exceed those required for passing the government approval process. Antle (1995) shows that this reduces the need for "command and control" type of government intervention. In some EU countries, economic sanctions are very limited in the case of food safety problems. In France, liability is limited, and in nonlethal food poisoning problems, plaintiffs seldom take legal action. When an unsafe product is marketed, resulting in the deaths of consumers, this most of the time results in penal sanctions for the manager rather than large economic sanctions for the firm (for example, to our knowledge, the fatal poisoning of hundreds of people with tainted cooking oil in Spain has not resulted in any significant monetary compensation for the plaintiffs 16 years later). That is, the incentive for supplying safe products would be perhaps less than in the United States, in the absence of a command and control regulation. Fundamental differences in the legal system for protecting consumers from health hazards provide some justification for diverging conceptions on the role of government in setting standards. More generally, differences in the legal environment, such as ex ante regulation versus ex post litigation as a basis for law, may provide justifications for differences in governmental standards between countries.

Cultural Differences

Arguably, the fact that Islamic countries tend to erect barriers to pork meat imports is not seen as an unfair nontariff barrier. One may wonder whether a ban on genetically modified organisms (GMOs) should be considered as legitimate when it has some religious connotations (Egypt announced such a ban in 1996, even though it actually never enforced it). This raises the question of how far one should go in this area and whether the concerns of consumers in Luxembourg and Austria on GMOs (which look genuine) can also legitiman an import ban. A recent survey measured consumer acceptance of GMOs in 19 countries and showed that only 22 percent of Austrian consumers seem willing to buy genetically modified products, against 74 percent in the United States (Hoban, 1997). Even in Britain, one of the most permissive countries in the EU in this area, a poll showed that only 14 percent of consumers were happy with the introduction of genetically modified foodstuffs, and 96 percent wanted labels on food made of genetically modified seeds (*The Economist*, 1998). This reluctance cannot be explained completely by a lack of information because Hoban's survey reported that a higher proportion of consumers than in the United States said that they had read or heard information about biotechnology. Part of the explanation seems to lie in cultural factors. Most of the consumers' concerns actually seem to be linked to the possible spread of unwanted genes in the environment, rather than concerns about their own health.

The case of GMOs is an illustration of the impact that consumers' cultural values can have on trade, regardless of scientific considerations. It is not the only one. Consumers in some countries remain opposed to irradiation, which is seldom used as a result (except for specific products such as spices, onions, and some poultry in certain countries), even though the International Atomic Energy Agency and the WHO have concluded since 1980 that irradiated food presents no toxicological risk. Dissimilar consumer preferences have an impact on trade, as for example if one country requires ground meats offered for sale on its territory to be irradiated and another refuses to use the technology. Here again, despite scientific considerations, even very subjective quality considerations can have an indirect effect on trade.

Ethical Concerns

Animal welfare regulations, introduced under pressure from animal rights activists, are becoming a very important topic in Europe. This may also have large consequences for international trade. In some cases, imports of products that do not comply with certain rules may be prohibited. The EU, for example, has banned imports of furs of 13 types of animals that can be caught using leghold traps, even though not all of them are listed by the International Convention on Trade in Endangered Species and some species are farmed commercially. National regulations may also distort competition. The EU has adopted directives banning the battery farming (i.e., rearing in individual boxes) of milk-fed veal calves and has imposed collective rearing, including cellulose feedstuffs, which considerably increases production costs. This measure, still in the transitional phase, has little impact on trade because there is little international trade in veal. Similar measures are being prepared for poultry, however, which could have a very considerable impact on the competitiveness of European poultry and egg producers, not only in export markets but also within the EU if third countries do not adopt similar measures. The planned increase in the size of cages (in egg production) and the possible animal density limitation (in poultry meat production) would cause a substantial rise in heating and feeding costs and hence in the cost of European poultry, in relation to countries not under the same obligations. Substantial trade flows could be affected.[7] Farmers, especially in England and Sweden, where the regulations are already stricter than in the rest of the EU, claim that consumer concerns should lead to similar requirements for imported products.

Growing numbers of consumers are also concerned about the possible adverse effects of their purchases, on the destruction of natural resources in other countries, for example, or on child labor (Mahé and Ortalo-Magné, 1998). There is growing pressure from public opinion for the imposition of more environment-friendly practices in third countries, especially to protect the "common resources of humanity" such as tropical rain forests. Some governments support a ban of timber products from countries where forests are threatened (Vogel, 1995). Consumers are also concerned about the importation of goods that they reject for cultural or religious reasons. But the fact that consumers' ethical values are not the same in all countries is bound to affect trade.

---

[7]Simulations with the MISS model of INRA and Ecole Nationale Supérieure Agronomique de Rennes (ENSAR) in France suggest that EU poultry meat exports could fall by 70 percent if the limitation of density asked by animal rights activists (16–18 chickens per square meter instead of the present density of around 23–26) was adopted. The competition of cheap U.S. chicken on the EU market would also increase dramatically.

Mistrust in Science

In some countries, there has been a growing mistrust in science over the past decade. France is a typical example where the government has minimized the effect of major accidents, which have fueled suspicion and eroded public confidence in science. The importance of asbestos-related cancers has been largely hidden under the pressure of the industry, and, when disclosed, past responsibility of mandated doctors in spreading wrong information has had a very negative effect on public opinion. Involvement of scientists in hiding information from the public in the nuclear sector has had a similar effect (scientists from government agencies claimed that the Chernobyl radioactive cloud has stopped exactly at the French border, something that nobody has actually believed). So has the continuous denying of public agencies involvement in spreading HIV-contaminated blood, until journalists disclosed evidence. In France, as well as in most EU countries, the poor management of information about possible links between bovine spongiform encephalopaty (BSE) and Creuzfeld-Jacob disease (CJD), and assurances from government-appointed scientists made the mistrust of science a very sensitive issue in the food sector. Consumer concerns about GMOs and growth activators cannot be understood without taking this into account. Although educating consumers is one of the government's task in this area, the situation is such that any government information is considered as suspect.

The poor management of information on past accidents by scientists and politicians is not the only reason for consumers' unease with science in some countries. One cannot blame only consumers' ignorance when they are not satisfied with the assurances given by biologists concerning the level of the *Codex* standards. Powell (1997) highlights the difficulties of obtaining reliable scientific assessments of the hazards present in food because of genetic mutations, for example, or combinations of pathogens with uncertain effects, or the influence of exogenous and unforeseeable factors on micro-organisms. The standards accepted by scientists do not always have an indisputable scientific foundation. Some have had to be completely revised at various times, and scientific "certainty" is sometimes fragile, especially with regard to the carcinogenic properties of products (Mazurek, 1996).

In addition, consumers assert the right to entertain fears that scientists regard as "irrational," especially concerning GMOs and irradiated food. Controlling short-term risks does not mean that long-term risks or uncertainties do not exist. In Europe, environmental and consumer groups have recently campaigned for the inclusion in multilateral agreements of a "precautionary principle," which would allow exceptions to the regulations in cases where scientific proof does not go far enough (Godard, 1997). This does not seem to be as much of an issue in the United States and in the rest of the world, except perhaps on some environment-related issues (Rege, 1994).

ACCOUNTING FOR CONSUMER CONCERNS

Consumers' Values Matter

Governments and international agencies should not dismiss consumer concerns about food safety, nor about ethical, environmental, or cultural values, and even perhaps about imagined health risks. This could significantly erode public support for the trade liberalization process. Mandatory compliance to ill-accepted standards may result in consumers' rejection of freer trade (Olson, 1998). The 1991 GATT panel on tuna, which basically ruled that a country could not ban imports for environmental reasons outside its territory, dragged into anti-GATT movements hundreds of thousands of people who would never have joined such organizations otherwise (*The Economist*, 1993). Threats from the European Commission to restrict the sale of unpasteurized cheese are said to be "responsible for 5 out of 6 French votes against Maastricht." The estimate quoted by Vogel (1995) may not be completely statistically exact, but it truly reflects how anti-EU populist groups exploited fears of being "condemned to eat standardized, aseptic, industrialized cheese." The 1997 panel on hormone treated beef also had a strong negative impact in European public opinion. Because of this panel, the WTO is now often perceived as an international agency whose goal is to overcome countries' rights to protect their consumers and more generally to undermine national sovereignty. Both issues contribute to fuel the isolationist propaganda of interest groups and extremist political parties (namely the far-right wing) in France and several other EU countries.

Clearly, ignoring consumers' concerns could lead to a severe rejection of "globalization," which already has a poor record in public opinion in many countries. Food is a sensitive topic, and few things are likelier to give trade liberalization a bad name than to have it associated with foisting on consumers the eating of mediocre or even potentially unsafe food. This aspect should not be underestimated, especially when consumer groups can find in the Internet a powerful soundboard. The international coordination of the opponents to the OECD's proposal on Multilateral Agreement on Investment shows that decisions can no longer be taken without consumers' approval (and more generally without citizens' approval). The list of Internet web pages gathering protests against the WTO, and even the SPS agreement, is becoming very large.

Even when they do not sign petitions, demonstrate, or protest in the polling booth, consumers can react by changing their consumption patterns. If consumers consider the way in which children are exploited, cosmetics tested, foxes killed, or cattle reared to be an integral component of the quality of a food product, lipstick, coat, or piece of meat, their demand for such products is altered by the presence or absence on the market of goods which do not comply with their ethical values. This is the case, for example, if it is difficult for consumers to identify goods produced under such conditions.

"Bad" Products Driving Out "Good" Products

Another reason for taking consumers' concerns into consideration is that there are some externalities[8] between unsafe (or nonpolitically correct) and safe (virtuous) products. When consumers are not able to distinguish the specific quality of different products, they are not willing to pay as high a price as they would if they were sure that the product was of high quality.[9] Akerlof (1970) has shown that imperfect consumer information about product quality could even result in total closedown of the market (absence of trade) if, because of a lack of information, buyers' willingness to pay was insufficient to cover production costs. If buyers' willingness to pay is less than the cost of producing high-quality goods, only low-quality goods (less costly to produce) are traded and high quality is frozen out of the market. Akerlof used secondhand cars as a famous example of poor quality chasing away high quality, but in the food sector too, the workings of the market may cause vendors to offer an inadequate level of quality or safety when information is imperfect.

Consumer goods may be divided into search, experience, and credence goods. A good is a search good when the consumer is capable of assessing its quality before buying it, an experience good when the consumer discovers the quality only after consuming it, and a credence good when the consumer never discovers the quality of the good (or does so only in the very long term). Many agrofood goods fall into the "credence" category (Caswell and Mojduska, 1996). This is the case, for example, when the "safety" component of quality or the nutritional composition of a product are at issue. It is also the case with the ethical, cultural, or environmental components of quality. The economic mechanisms at work in these three categories are different. With experience goods, for example, the incentives for quality fraud are limited by consumer

---

[8]Externality occurs when actions of an individual or a firm affect other individuals of firms. There are two types of externalities (1) negative, where one firm imposes a cost on other firms but does not compensate them (for example, by polluting water that other firms use), and (2) positivie, where one firm confers a benefit on other firms but does not reap a reward for providing it (for example, the owner of an apple orchard may confer a positive externality on a neighboring beekeeper). (Definition adapted from J. Stiglitz, 2000.)

[9]The difference between what an individual is willing to pay and what has to pay for a product is called "consumer surplus." This difference also represents a monetary evaluation of the welfare that is provided by the acquisition of a particular good or service. For example, if a person is willing to pay $50 for the first shirt, $45 for the second shirt, $40 for the third, etc., but the market price is $29 per shirt, this person gets a surplus of $50 – $29 = $21 for the first shirt, $16 for the second shirt, etc. If this person is willing to purchase the fifth shirt for $29, he or she gets no extra "utility" for this last shirt, since the price is equal to the willingness to pay. The fact of being able to access a market where shirts cost $29 provide a surplus of $21 + $16 + etc. for the previous shirts, which is a measure of a person's welfare brought about by the availability of shirts at this particular price. Note that under the combined effect of a lower price and extra consumption this welfare increases only if shirts cost less that $29.

sanctions on the occasion of repeat purchases. With credence goods, there is no spontaneous mechanism for market regulation and it is more difficult to indicate quality in a credible way. The market failures highlighted by Akerlof may extend into the long term.

Opening up markets can result in the coexistence alongside local products of foreign products whose quality is less familiar to domestic consumers. The imported goods may be perceived as being of lower quality because of doubts as to foreign control procedures or the different importance attached to each component of the overall quality of the good. Consumer uncertainty as to the type of products on the market (which might result, for example, from imports of goods such as hormone-treated meat or genetically modified seeds) could affect demand, decrease consumers' willingness to pay, and raise adverse selection problem (Bureau et al., 1998). It is theoretically possible for the welfare loss resulting from reduced consumer willingness to pay to outweigh the welfare gain resulting from cheaper imports. For example, the European Commission suggested that lifting the ban on imports of hormone-treated beef, right after the mad cow crisis, could lead to a 20 percent decrease in beef consumption in the EU (Hanrahan, 1997). We are not aware of any rigorous study that supports this particular figure, which clearly seems to be an upper bound. However, most observers agree that this would cause an extra decrease in consumption in this market. Although the magnitude of the fall in willingness to pay is difficult to assess, it is possible that the losses for the European economy could be large, in comparison to the gains for U.S. exporters. In such a case, one may consider that opening preferential access quotas to U.S. meat from certified producers who do not use hormones might be a better solution, from the viewpoint of overall welfare.

WHAT ARE THE SOLUTIONS FOR RECONCILING CONSUMER CONCERNS AND INTERNATIONAL TRADE RULES?

Consumer Concerns in the United States and Europe

Different consumer organizations have expressed their displeasure with international trade rules about food safety and quality. Several U.S. organizations, such as the Center for Science in the Public Interest, Safe Tables Our Priority, or Public Citizen, Inc., have complained that *Codex* standards are less protective of consumers than some domestic standards. These organizations are concerned that the SPS Agreement, because it facilitates trade and (allegedly) results in a "downward harmonization" of health and safety standards, could contribute to an increase in pathogen outbreaks (Fox, 1998; Public Citizen Global Trade Watch, 1998). Some of these groups are, for example, pressing for a revision of Article 10 of the SPS Agreement, which recommends taking account of the special needs of developing countries in the definition of standards (Silverglade, 1998). It is worth noting that the food safety issues raised by such organizations seem to have played a role in the refusal of the "fast-track" negotiation procedure by the U.S. Congress. During the 1998 debate on the Safe Food Act, government representatives announced on many occasions that they would reinforce inspection procedures and that imports would meet stricter controls, showing that consumers' protests are increasingly affecting international trade arrangements.

In Europe, a major concern is that standards should reflect what consumers want in a product, not what the industry wants to put into it. That is, European consumer concerns include more and more cultural and environmental attributes of quality. Organizations such as Greenpeace argue that WTO rules should be amended in order to cope with consumers' values, and such views meet strong support in some EU countries. Consumers and environmental groups have been rather successful in lobbying the European parliament in some areas. Bowing to the pressure of public opinion, the European Commission as well as some national governments have introduced regulations in areas such as animal welfare, the protection of fauna and flora, and GMOs. It is noteworthy that some of these regulations considerably increase the cost of producing food. As a result, these governments point out that the SPS and TBT agreements put them in an uncomfortable position by forcing them to authorize imports of goods produced using methods which they have had to ban at home. The EU is unhappy at being obliged to authorize imports of food produced under less restrictive livestock farming conditions than its own (e.g., animal welfare), or using biotechnologies that consumer pressure prevents its own farmers from using (e.g. bovine somatotropin [BST or rBGH], a hormone used to increase milk yields in some countries).

Although one cannot ignore consumer concerns, fears about food safety give trade protectionists a wonderful opportunity to cheat; and trade restrictions motivated by social, cultural, ethical, or environmental considerations can be a form of protectionism in disguise. There is often convergence between consumer demands for stricter standards than those recommended by scientists and the economic attraction of strengthening nontariff barriers. The true motives of a government saying that it is barring imports in order to stop people feasting on unclean fowl, or on meat stuffed with synthetic growth hormones, or maize that has been modified by frightening new technologies (each case being the subject of a quarrel between the United States and Europe) are hard to discern (*The Economist*, 1997). Because this problem is likely to be a major area of contention in coming years, it is worth exploring possible ways to address it.

More "Sound Science" in the SPS Agreement

After the Uruguay Round, there were large hopes that the reference to "sound science," and in practice, the provisions of the SPS Agreement that make a clear reference to international standards, would solve any potential conflict on SPS issues. However, things have proved more complex in practice.

The idea of objective science serving to guide trade practice, which prevails in the SPS Agreement, is debatable. In practice, economic and political considerations are very much intermingled. In many cases thresholds have been set not only on the basis of medical effects but also on the basis of what is technically and economically feasible, and many scientists acknowledge off the record that some standards are defined "after the event" (radioactivity thresholds, for example). Ever since scientists' recommendations acquired the status of potentially mandatory standards, with considerable economic interests at stake, it has been difficult for them to ignore economic considerations. Salter (1988), Powell (1997), and Hillman (1997) have given numerous examples of "mandated science" or "negotiated science." Manufacturers are also strongly represented on *Codex* and joint FAO/WHO committees, and economic interactions with standard setting are obvious.

More generally, a trade-off between costs and benefits is sometimes implicit behind the scientific criteria in the form of the setting of standards, which take economic factors into consideration and reference to risk analysis in the settlement of disputes. Risk analysis includes a risk management component; this corresponds to the ways in which risk may be reduced to an "acceptable" level, which includes economic considerations, and in the last resort the decisions taken are often of a political nature.

Science is not always completely conclusive. Many scientists express their doubts about the way standards for chemical residues are defined (according to Antle, 1995, U.S. Environmental Protection Agency estimates for cancer risk from pesticide residues are approximately 1,000 times higher than equivalent risk estimates using other methods). In many cases, standards have been established on the basis of experiments on mice and rats and extrapolated to humans. Even with a considerable safety margin, the basis for such standards seems relatively arbitrary. (In France, for example, some scientists recently disclosed how fragile the basis is for defining standards on dioxin, which may be found in dairy products and accumulates in organisms and in mothers' milk when breast-feeding). International standards are now put to the vote at the *Codex*, and some are passed by a small majority. Not all countries are willing to acknowledge the legitimacy of risk thresholds imposed on them in this way.

"Sound science" and the reference to "available scientific evidence" in the SPS Agreement may in practice conflict with the precautionary principle, which is more and more referred to by consumers' organizations. This problem is of particular importance because it has recently led to a very controversial situation on the issue of GMOs in the EU. According to the precautionary principle, precautionary measures should be taken in this absence of certainty according to the state of scientific knowledge at the time. Although it is not a legal principle, it can be reflected in regulations.[10] The preliminary decision of the French

---

[10]The precautionary principle is recognized in several international agreements (e.g., International Convention on the Protection of the North Sea, Rio Declaration, Framework Convention on Climate Change), in European law (Maastricht Treaty), and in national law (French 1995 law on environment, U.S. law on pharmaceutical approval). However, in many cases, European consumers' organizations give a much broader scope to this principle than the somewhat restricted version mentioned in international agreements. The Rio Declaration for example, only states that "in order to protect the environment, the precautionary approach shall be applied by a State, according to its

Conseil d'Etat (a kind of supreme court) in September 1998, in favor of nongovernmental organizations which asked for a ban on GMOs, partly relied on the fact that the precautionary principle was embedded in the 1995 environmental law.[11] Article 5.7 of the SPS Agreement indicates that, if relevant scientific evidence is "insufficient," members may adopt SPS measures, on a provisional basis, while seeking additional information about the risks posed by a recently identified hazard. However, provisions of the SPS Agreement regarding precaution are much more restrictive than what some consumer groups often mean when they invoke the "precautionary principle," suggesting that there may be a fundamental ambiguity between the expectations of certain groups in society and practical measures.

For all these reasons, the reference to sound science is not the panacea that is often described by international organizations. Other ways of reconciling consumer concerns should be taken into consideration.

Labeling

Many economists see labeling as an efficient way to solve disagreements about harmonization. The idea is that one should "give consumers the choice." Beales et al. (1981) have shown that segmenting the market, and allowing for each group of consumers to buy the products corresponding to their willingness to pay, is, in theory, a much better solution than mandatory uniform standards. As a result, labeling and consumer information policies are often portrayed by international organizations as preferable alternatives to regulation because they are cheaper for producers, leave the choice to consumers, and are less likely to constitute trade barriers (OECD, 1997b).

However, in practice, labeling does not solve all problems either. First, labeling is not always possible, or, when it is, it can be very expensive. The proposals for a strict labeling of GMOs in Europe require complete traceability, that is, that the whole chain be segmented, from the producer to the final processed product. According to the industry, this would generate very large costs (a Canadian study into segregating modified wheat products found that this would require separate facilities at 15 different points from farm to market). In addition, pollen is known for spreading between controlled and uncontrolled areas (some pollen was found at several kilometers in altitude), and the

---

capabilities. Where there are threats of serious or irreversible damages, lack of full scientific certainty shall not be used as a reason for postponing cost effective measures to prevent environmental degradation" (Principle 15 of the Declaration of the United Nations Conference on the Environment and Development, adopted in 1992).

[11]In December 1998, the Conseil d'Etat postponed the final decision because the legal consequences of this principle were unclear and contradicted other legal texts (i.e., the EU Directive 90/220 which regulates the approval of deliberate releases of GMOs into the environment). It temporarily upheld its preliminary ban on Novartis' genetically engineered maize and asked the European Court of Justice to give its advice on the legality of the initial approval.

segmentation of the two markets can hardly be perfect. In other cases, labeling is simply not the solution that consumers are willing to accept. For example, animal welfare activists have clearly stated that labeling is not an issue and that they want an interdiction of certain rearing practices.

Another reason why labeling is not the panacea described by some economists is that the conflicts about the appropriate level of standards are sometimes simply displaced toward the issue of the appropriate label, which is equally complex. There are diverging opinions, for example, on the relevance of labels on clothing certifying low levels of pesticide use in the production of cotton, or on the specifications for labels certifying that wood products do not harm tropical rain forests. Mutual recognition of labeling for organically farmed products is difficult to achieve because countries apply the relevant criteria more or less strictly, or because some countries are considering granting such labels to genetically engineered or irradiated products. In November 1998, this was a bone of contention between EU member countries, and the adoption of an EU-wide definition of "organic" ("biologique" in French) food was postponed. Basically, the need for international harmonization and recognition of labels and of the underlying certification procedures raises difficulties that are comparable to the ones raised by the harmonization and recognition of mandatory standards

Finally, economic theory suggests that, if agents are rational, a label on credence goods should not be sustainable. The idea is that rational consumers know that they cannot verify that producers fill their commitments, while rational producers have no reason to do so. Labels on credence goods require a third-party certification, and, in spite of that, are not always trusted by consumers. In particular, it is difficult to monitor the production process of imported credence goods, which is the sole means for acquiring information about their quality. Foreign firms are also less exposed to judicial sanctions (liability), which may encourage fraud when the consumer is unable to verify the quality of the good in question directly.

Expanding GATT Criteria

Some consumer values may well be out of step with GATT principles. In the environmental sphere, rulings in disputes brought within the framework of GATT and the WTO hold that a country is not entitled to use trade measures restricting imports to protect natural resources outside its territory, even in the case of resources that some consider to be "common to humanity." In the cultural sphere, a country may introduce regulations that are more stringent than international standards on ethical, moral, or religious grounds under only very limited conditions.[12] Recently, the 1998 Appellate Body on the shrimp–turtle case ruled that Article XX exceptions are "limited and conditional," a confirmation of the first GATT tuna panel conclusion, but which could prove important in future challenges to domestic health, safety, and environmental

---

[12]Article XX(a) allows import restrictions when they are "necessary for the protection of public morals."

regulations. The SPS Agreement does not recognize the validity of consumer concerns in cultural, ethical, and environmental areas (although under the TBT Agreement they may be taken into consideration by authorizing different labeling). Because of the mismatch of GATT rules and some consumers' concerns, many environmentalist and consumerist organizations claim that one should include other factors than SPS risk in the *Codex* and the SPS Agreement, or that one should give a broader scope to Article XX so that it embeds a larger set of consumer values.

A number of arguments can be made for including ethical and cultural values as grounds for trade restrictions. Reluctance to consume goods produced in unethical conditions can affect demand for all goods, including those produced "virtuously," as for example with animals caught in traps and animals reared on farms. Externalities between goods may arise if ethical and cultural values are acknowledged. Trade is also one of the most effective means for obliging countries to respect human or children's rights or to protect natural resources and endangered species. Socially aware consumers who would like to be able to wield such a weapon find it hard to understand why international trade rules should prevent them from doing so. Vogel (1995) gives several examples where trade restrictions and a desire for access to greener markets have had an impact on a country's attempts to improve social and environmental regulation. Moreover, it may be paradoxical to reject trade restrictions for cultural reasons when they are admitted for nonfood products such as medicines. The case of RU 486, the "morning after" contraceptive pill, is an extreme example of a product that may not be imported into certain countries, including the United States, solely for cultural reasons in spite of evidence showing large reductions in hazards for women, linked to pregnancy interruption at a later stage. Last, thresholds and standards are sometimes adopted in line with what is socially acceptable, and reference to an acceptable risk introduces cultural considerations into the SPS Agreement (the acceptability of a given risk is subjective although required to be scientifically justified if different from international standards), raising the question of whether this type of consideration should be included explicitly in the agreements.

However, giving consideration to ethical, cultural, or moral arguments could open a Pandora's box. For some countries, risk may be social as well as biological, including factors such as bankruptcy among farmers and rural desertification. Cultural or ethical arguments could be used to cover a potentially unlimited number of exceptions to free trade. A lax interpretation of the TBT Agreement in this sphere would provide justification for a whole host of trade barriers. In practice, this debate has already been raised within the *Codex*. The legitimacy of socioeconomic and cultural factors has been a bone of contention for years in *Codex* committees, namely on the issue of BST. Some countries argued that economic and social factors and consumer reluctance should be taken into account. But if a decision has been taken at the *Codex* to defer and reconsider the BST case, it is not on account of these arguments.[13]

More generally, an agreement on the consideration that should be given to arguments other than "objective" medical risk when sanitary regulations are being defined seems difficult to reach. Discussions on the item on "the role of science and the extent to which other factors are taken into account" at the 13th Session of the *Codex* Committee on General Principles (September 1998, Paris) have been largely inconclusive. A similar problem exists within the dispute settlement procedure of the WTO. Decisions taken on the basis of purely scientific considerations simply seem unacceptable to consumer organizations who expect that international standards reflect what they want to eat, not only what is safe to eat. When WTO rules conflict with decisions of a democratically elected parliament (as in the EU and U.S. hormone-treated beef issue), things get even more difficult.

Can Economics Help?

When cultures differ, economic analysis may perhaps help in finding a common playing field. This approach progressively has been accepted in the area of environmental disputes, and it is progressing, albeit slowly, in the phytosanitary area, and, to a lesser extent, in the sanitary area. Here we consider the possibility of a broader use of economic assessment in food quality regulations as well as in dispute settlement on nontariff barriers.

Cost–benefit analysis is already used to enable public authorities to make decisions concerning national regulations. It is an important stage in the framing of regulations in the United States. Arrow et al. (1996) recommend that the method should be used systematically because they observed considerable differences between the cost of public health measures and their real impact on health (they give estimates where, within the same agency, the cost per life saved varies between $200,000 and $10,000,000 [U.S] depending on the program, which means that more lives could be saved at the same cost to society; see also Magat et al., 1986). Even though society does not accept all risks in the same way, and even though social choices cannot be reduced to the

---

[13]In 1990, the EU imposed a moratorium on the use of BST until the end of 1999 (although without banning imports of dairy products from countries where BST is allowed). The *Codex Alimentarius* approached the problem of BST and growth hormones from the standpoint not of farming practice but of measurable residues, which proved to be low in both cases. The consumer representative and several countries argued that consumers were opposed to the use of BST and that BST improved neither the quality nor the health characteristics of milk, and asked to be allowed to ban it. The EU asked for "legitimate factors other than scientific analysis" to be taken into consideration. But the vote to defer the decision was taken because some delegations had contributed scientific evidence which raised questions about the weakening of the immune systems of animals treated with BST and argued that this could increase the risk of infection, the need for treatment, and hence levels of antibiotic residues.

equalization of a statistical cost between programs, cost–benefit analysis should take a more important place.

Box 7-1 describes a few possible techniques that can introduce more economic assessment into the SPS regulations. There are clearly many technical difficulties. Measuring the benefits procured by regulations designed to guarantee certain ethical or cultural aspects of product quality is no easy matter, and the problem of the valuation of imagined risks is a difficult one (Pollak, 1995, 1998). Estimates of cancer risk from pesticide residues contain a substantial degree of uncertainty as to the risk, making any economic estimate particularly difficult. When cost–benefit analysis builds on the imprecise data of risk analysis, results are often subject to very large confidence intervals. In addition, it is not possible to calculate the probability of a risk that is too uncertain, making it difficult to carry out analysis with conventional tools. This is the case, for example, with the risk of GMOs propagating genes, or the risk of long-term epidemics such as BSE and CJD, or environmental risks. However, similar problems exist in traditional risk assessment procedures. With an economic approach, it is possible to use approaches based on the measurement of changes in the consumer utility function when consumers have access to a product with attributes to which they are attached (Kopp et al., 1997).

When human health is at stake, the topic is more sensitive because giving a value to illness avoided or even a human life saved is not always well accepted, especially in some EU countries. However, it is worth noting that in the same countries, transportation and energy departments use such calculations on a daily basis when they decide priority investment in road safety or thresholds in building dams. Economic assessment would simply make choices more explicit, although concepts such as "the value of life" (actually, the value of life saved) can still be shocking for many people (Viscusi, 1993).

Cost–benefit analysis can be of particular interest as far as ethical or cultural values are concerned. If, for example, consumers place particular value on the fact that a good is produced without the use of biotechnology or irradiation techniques, estimating their willingness to pay means that the variation in consumer satisfaction resulting from a regulation prohibiting the technique in question can be quantified in money terms (Viscusi et al., 1995; Magat and Viscusi, 1993). One application could be the animal welfare issue, an awkward case where public opinion is being represented by vociferous consumer lobbies in Europe, and where scientists have proved to be of little help. More economic assessment would make it possible to assess the real importance of this concern throughout the entire population.[14]

---

[14]The French Institut National de la Recherche Agronomique (INRA) has recently started several academic studies involving either contingent valuation (the measurement of willingness to pay for guaranteed prion-free meat, for "animal-welfare-correct" food at INRA-Rennes) or experimental economics (organic food, GMO-free products, etc., at INRA-Ivry).

BOX 7-1.
Methods for Estimating the Benefits of Sanitary and Technical Regulations

Where food safety and the spread of plant and animal diseases are concerned, cost-benefit analysis involves quantifying the level of risk and estimating its economic impact. This approach is widely used, although very unevenly from one country to another, not only in order to assess the interest of a regulation but also to compare the advantages and disadvantages of several possible means of government intervention. In particular, it can be used to rationalize the strengthening of SPS controls in relation to the dissemination of information and the raising of consumer awareness, or to inform decisions about the introduction of regulatory standards (Kopp et al., 1997).

Although there are still some technical difficulties, there are few major obstacles (except the lack of an economics culture among the administrations in some countries such as France) to complementing classical risk analysis by cost-benefit analysis in the phytosanitary and animal health area. Things are more complex, however, when cultural values are at stake and when one deals with human health issues.

Several methods exist for estimating the cost of mortality and morbidity and evaluating in money terms the benefits of government action resulting in a reduction of sanitary risk. With the human capital method, a value is placed on the reduced risk of premature death based on an evaluation of discounted labor flow. For an individual of a given age, the value of the life prolonged (statistically) by a regulation corresponds to the discounted sum of the mathematical expectation of the person's income (Freeman, 1993). Some extensions of this method have been proposed, in particular by integrating nonmonetary aspects and the value of the individual's descendants (Viscusi, 1993). With the cost of illness method, a value is placed on the reduced morbidity resulting from sanitary or regulatory methods, based on an estimate of medical costs and productivity losses due to illness (Buzby et al., 1996; Crutchfield et al., 1997). Opportunity costs from investing in activities that reduce the risk are included in the value of reduced illness (Landelfeld and Seskin, 1982). As with the human capital method, statistical methods have to be used to estimate the risk, especially dose-effect relationships.

Methods based on estimates of willingness to pay, although more difficult to apply, are wider reaching because they make it possible to include quality-related aspects that cannot be translated into identifiable short-term illness. The preventive expenditure method seeks to measure agents' willingness to pay by observing the efforts made to avoid illness. With this method, a money evaluation of the disutility of being ill is added to the estimated cost of illness, together with an estimate of the preventive expenditure that an individual is willing to commit according to a given pathogen level (Harrington and Portney, 1987). Contingent evaluation methods involve asking individuals directly about their willingness to pay in order to reduce the risk of an illness, or more generally to obtain higher quality in a good. By directly revealing willingness to pay, this method theoretically makes it possible to gain a money estimate of all the benefits arising from a given measure. However, answers have to be corrected for statistical bias due to respondents' incentives to over- or underestimate their willingness to pay (which depends in particular on whether they anticipate having to pay the disclosed sum or not). As these methods are widely applied to environmental issues, efforts have been made recently to harmonize survey methodologies (see, in the United States, National Oceanic and Atmospheric Administration [NOAA] Panel on Contingent Valuation [Arrow et al., 1993]). Another method being used increasingly widely at present is the experimental economics method, which involves getting a group of individuals in a situation where their real behavior is simulated to reveal their willingness to pay for particular qualities. Such methods are relatively onerous to put in place, but they make it possible to obtain a precise measurement of the value that a sample of individuals places on different sanitary thresholds, according to information received, for example (Hayes et al., 1995).

The methods described above are used to evaluate the benefits of drawing up a regulation to protect consumers' health or to ensure that they acquire the quality they desire. Methods for evaluating the cost of regulations are generally based on estimates of the welfare loss of the agents concerned when they have to comply with a regulation. This includes, for example, the cost to firms of acquiring suitable equipment and many other direct and indirect costs. Kopp et al. (1997) provide illustrations of such estimates. One method involves valuing them as opportunities that had to be foregone. This includes the diversion of resources, the value of specific inputs that become useless, the excess cost of substitution technologies, and the price differentials with replacement products borne by the consumer.

Source: Adapted from OECD, 1999.

ECONOMIC ANALYSIS AND THE SETTLEMENT OF DISPUTES

Scientific and Economic Criteria

Present arrangements for the settlement of international disputes relating to technical and sanitary barriers have put economic analysis second to risk assessment. The SPS Agreement recognizes that governments may set higher sanitary standards than the ones used in other member countries. In practice, this means that one can restrict imports on sanitary grounds, when a hazard is scientifically proved to exist, and that one cannot implement such import restrictions in the absence of proof of significant hazard. Although Article 5.3 of the SPS Agreement (and Article 2.2 of the TBT Agreement) mentions economic assessment, such considerations have only a limited place in the settlement of sanitary and technical disputes, and cost–benefit analysis is far less central than risk analysis.

International agreements on sanitary and technical measures do not oblige countries to adopt only those regulations whose benefits exceed their costs (Roberts, 1997). In practice, many countries introduce import restrictions on sanitary grounds, to avoid the spread of pests for example, without making any prior estimate of potential losses. These may sometimes be very small in comparison with the cost to consumers caused by the regulation in question. If economic methods of calculation were used more systematically, the welfare gains resulting from the import restrictions could be compared with the welfare gains resulting from freer trade (James and Anderson, 1998; OECD, 1997b).

Assessing Consumers' Concerns

In addition to helping decision makers when choosing between different risk management options and when reviewing quarantine policies, cost–benefit analysis can also provide a sounder basis for discussing the role of "other legitimate factors" than health hazards, a problem that remains a live issue.

It is often argued that *Codex* standards and texts express policy choices and that such policy choices could extend to national policies in such areas as the environment, consumer concerns, animal welfare, and societal values. If these values were considered from an economic standpoint, the debate might lead to more convergence in the different points of view. Because it is based on a revelation of individual preferences, cost–benefit analysis can be seen as a tool for organizing many different pieces of information and points of views in a consistent framework.

Up to a certain extent, willingness to pay is a defendable measure of people's concerns. Genuine consumer aversion, for sanitary as well as for cultural reasons, is reflected in a willingness to pay in order to avoid the products. Although there are still some technical difficulties and conceptual obstacles, contingent valuation techniques or experimental economics may help people from different cultures to find a common "metric" for defining more objectively how genuine the concerns of their consumers are, and for finding solutions to complex issues which largely reflect cultural differences. In this respect, microeconomics can be seen as a useful negotiation language.

More Economics in the Settlement of Disputes?

The procedure for settling sanitary and technical disputes under the auspices of the WTO could draw on the experience of competition policy. One accepted principle, including in international disputes, in competition policy is that certain forms of coordination between producers, which may indeed restrict competition, are not necessarily undesirable from a social standpoint. In most developed countries, public regulators (competition councils, antitrust commissions, etc.) weigh up their advantages and disadvantages (Viscusi et al., 1995). Infringements of competition rules are permitted after an economic cost–benefit analysis, and noncompetitive arrangements are often accepted if it can be proved that they bring economic benefits and that the benefits are fairly distributed between agents. Regulators tend to make such decisions on a case-by-case basis, weighing the pros and cons and carrying out a mainly economic cost–benefit analysis rather than applying immutable general principles. Less consideration is given to such principles in the settlement of sanitary and technical disputes, especially in the international arena.

All regulatory measures likely to hinder imports are sometimes classified as nontariff barriers. Some particular studies in the agrofood sector use a broad definition of the term nontariff trade barriers (Hillman, 1991; Roberts and DeRemer, 1997). However, Baldwin (1970) has suggested that nontariff barriers should be defined as policies that reduce potential world revenue. According to this definition, policies, which in practice restrict trade flows, would not be regarded as nontariff barriers if their effect was to correct market inefficiencies and increase world revenue. Mahé (1997) proposes extending the definition to include nonmonetary effects. He suggests that measures whose elimination would cause welfare losses in some countries that are greater than welfare gains in other countries should be classified as nontariff barriers. This definition is in line with both economic theory and the idea of using cost–benefit analysis to arbitrate disputes.

When trade liberalization calls into question national regulations whose effect is also to reduce market inefficiencies, the welfare effects may be analytically ambiguous (Thilmany and Barrett, 1997; Bureau et al., 1998). If a WTO panel, for example, results in an obligation to import products that do not satisfy consumers' ethical, environmental, or cultural concerns, antiselection mechanisms could cause substantial welfare losses. In practice, this could involve consumer boycotts or rejections, which would affect demand for all the goods concerned, both imported and domestic. Estimating overall costs and benefits would involve quantifying the different variations in welfare, raising awkward technical problems. Nonetheless, it is possible for welfare losses to be greater than welfare gains at a global level. It would be paradoxical if trade liberalization, introduced by an international organization in the framework of the settlement of disputes, were to result in more trade but less welfare. In such cases, Baldwin's criterion could serve as a basis for settling disputes (Mahé, 1997). Practical implementation could be based on a cost–benefit analysis which would seem to be more in line with the maximization of collective welfare than are rigid principles derived from uniform scientific standards.

CONCLUSIONS

Many Americans may wonder why Europeans chose to ban hormone-treated beef and not tobacco, a far more hazardous substance. Yet French consumers find it difficult to understand why Americans support a ban on camembert cheese (made from fresh raw milk with the addition of a fungus, *Penicillium camemberti*), while they tolerate the risks linked to legal possession of handguns. Both American and French consumers will nevertheless look aghast at the Japanese who willingly pay an extraordinarily high price for eating the dangerous *"fugu"* fish, which regularly leads to death. Coping with such differences in the perception of risk within a uniform international code of standards and a "one size fits all" SPS Agreement is bound to raise a lot of difficulties and frustrations. The issue is even more complicated when attributes of food quality other than safety are involved (i.e., cultural, environmental, or ethical values).

Since the 1994 SPS Agreement, the reference to "sound science", has helped make legislation more consistent across countries. However, it is unlikely to solve all the problems. Science is not always conclusive, scientists' recommendations are not always trusted nor well accepted by consumers, and scientific risk assessment does not make it possible to account for the genuine concerns of consumers on other aspects than health risk (i.e., cultural, environmental, and ethical concerns) which are becoming a major area of contention in international trade (Baghwati and Hudec, 1996). In any case, measuring the risk gives no indication of the loss of utility for consumers. When focusing on risk analysis as the SPS Agreement presently does, one may run into the problem that although there is a very slight risk that a product is dangerous, the mere fact of knowing this to be the case could result in a very high proportion of consumers refusing to buy the product, and therefore high welfare losses (Josling, 1998).

Accounting for consumers' values, including factors other than health risks, could prove necessary in the future, if one wishes to avoid weakening support for trade. Food is a sensitive issue and free trade will be given a bad name if it is associated with the forcing on consumers of unwanted food. Genuine consumer aversion for certain imported products, for sanitary or cultural reasons, is normally reflected in a willingness to pay in order to buy other goods which satisfy their concerns. Giving this willingness to pay greater importance in the settlement of disputes, by comparing it with the costs to other economic agents, would help take account of consumer preferences. This could also help to prevent detractors of a more open trading environment from linking trade liberalization with an obligation to consume products that do not correspond to consumers' aspirations.

Economic analysis raises a number of technical difficulties. There is also the question of which version of cost–benefit analysis is the right one for the problem at hand, since economists are hardly of one mind on this issue. However, the methodologies described in Box 7-1 have raised similar difficulties in the evaluation of environmental costs and benefits, although agreements on evaluation procedures have progressed. What was considered as not feasible 20 years ago (e.g., the use of contingent valuation for assessing environmental damages and calculating fines in a trial) is now widely accepted. One may think that economic analysis in the SPS area is at a stage comparable to that of economics in the environmental area two decades ago. In many cases, cost–benefit analysis can already be a useful negotiation tool.[15] It will not solve everything, but given its potential contribution in the settlement of disputes, it deserves a more important role in the SPS area.

REFERENCES

Akerlof, G. 1970. The market for lemons: qualitative uncertainty and the market mechanism. Quarterly Journal of Economics 84(1):488–500.

Antle, J.M. 1995. Choice and Efficiency in Food Safety Policy. Washington D.C.: AEI Press American Enterprise Institute.

Arrow, K.J., R. Solow, P. Portney, E.E. Leamer, R. Radner, and H. Schuman. 1993. Report of the NOAA Panel on Contingent Valuation. Federal Register, 58(10):4602–4614.

---

[15]The economist Claude Henry defended such a position some 15 years ago in environment-related issues. On the basis of examples in the United Kingdom, he showed how cost-benefit analysis could be a constructive language for expressing public concerns. His paper was an important step in convincing environmentalists that serious economic evaluation could be in their interest, and the public in general that economics was not the evil science that many believed it was (Henry, 1984; we do not know whether this seminal paper has ever been published in English).

Arrow, K.J., M.L. Cropper, G.C. Eads, R.W. Hahn, L.B. Lave, R.G. Noll, P.R. Portney, M. Russell, R. Schmalensee, V.K. Smith, and R.N. Stavins. 1996. Is there a role for benefit–cost analysis in environmental, health and safety regulation? Science 272:221–222.

Baghwati, J.N. and R.E. Hudec. 1997. Fair Trade and Harmonization, Prerequisites for Free Trade? Cambridge, Mass.: Massachusetts Institute of Technology Press.

Baldwin, R.E. 1970. Non-Tariff Distortions in International Trade. Washington, D.C.: The Brookings Institute.

Beales, H., R. Craswell, and S. Salop. 1981. The efficient regulation of consumer information. Journal of Law and Economics 24:491–544.

Bureau, J.C., S. Marette, and A. Schiavina. 1998. Non-tariff trade barriers and consumers' information: the case of EU–US trade dispute on beef. European Review of Agricultural Economics 25(4):435–460.

Buzby, J.C., T. Toberts, C.J. Lin, and J. Macdonald. 1996. Bacterial Foodborne Disease: Medical Costs and Productivity Losses. Economic Research Service, Agricultural Economic Report 741. Washington, D.C.: U.S. Department of Agriculture.

Caswell, J.A. and E.M. Mojduska. 1996. Using informational labeling to influence the market for quality in food products. American Journal of Agricultural Economics 78:1248–1253.

Chen, J. 1996. A sober second look at appellations of origin: how the United States will crash France's wine and cheese party. Minnesota Journal of Global Trade 5:35–43.

Crutchfield, S., J.C. Buzby, T. Roberts, M. Ollinger, and C.T.J. Lin. 1997. An Economic Assessment of Food Safety Regulations: the New Approach to Meat and Poultry Inspection. Economic Research Service, Agricultural Economic Report 755. Washington, D.C.: U.S. Department of Agriculture.

Doussin, J.P. 1995. Le Codex Alimentarius à l'heure de l'Organisation mondiale du commerce. Annales des Falsifications de l'Expertise Chimique et Toxicologique 933:281–292.

The Economist. 1993. The Greening of Protectionism. February 27, p. 12.

The Economist. 1997. Fare Trade. May 17, p. 20.

The Economist. 1998. Food Fights. June 13, pp. 99–100.

Fox, N. 1998. Spoiled: the Dangerous Truth about a Food Chain Gone Haywire. New York: Basic Books.

Freeman, A.M. 1993. Measuring Environmental and Resource Values. Theory and Methods. Washington, D.C.: Resources for the Future.

Godard, O., ed. 1997. Le principe de précaution dans la conduite des affaires humaines. Paris: Éditions de la Maison des Sciences de l'Homme.

Hanrahan, C. 1997. The European Union's ban on hormone-treated meat. Report for Congress. Washington, D.C.: Congressional Research Service. June 5.

Harrington, W., and P.R. Portney. 1987. Valuing the Benefits of Health and Safety Regulations. Journal of Urban Economics 22(1):101–112.

Hayes, D., J. Shorgren, S. Shin, and J. Kliebenstein. 1995. Valuing Food Safety in Experimental Auctions Markets. American Journal of Agricultural Economics 77(1):40–53.

Henry, C. 1984. La microéconomie comme langage et enjeu de négociations. Revue Economique 35:177–187.

Hillman, J. 1991. Technical Barriers to Agricultural Trade. Boulder, Colo.: Westview Press.

Hillman, J. 1997. Nontariff agricultural trade barriers revisited. In Understanding Technical Barriers to Agricultural Trade, D. Orden, and D. Roberts, eds. St. Paul: The International Agricultural Trade Research Consortium, University of Minnesota.

Hoban, T.J. 1997. Consumer acceptance of biotechnology: an international perspective. Nature Biotechnology 15:232–235.

Hooker, N.H., and J.A. Caswell. 1996. Voluntary and mandatory quality management systems in food processing. Working Paper, University of Massachusetts, Ahmerst.

James, S., and K. Anderson. 1998. On the need for more economic assessment of Quarantine/SPS policies. Centre for International Economic Studies. Seminar Paper 98-02, University of Adelaide, Australia.

Josling, T. 1998. EU-US trade conflicts over food safety legislation: an economist's viewpoint on legal stress points that will concern the industry. Paper presented at the Forum for U.S.-EU Legal-economic Affairs, Helsinki, Sept. 16–19.

Kopp, R.J., A.J. Krupnick, and M. Toman. 1997. Cost Benefit Analysis and Regulatory Reform: An Assessment of the Science and the Art. Discussion Paper 97-19. Washington D.C.: Resources for the Future.

Landelfeld, J.S., and E.P. Seskin. 1982. the economic value of life: linking theory and practice. American Journal of Public Health 6:555–566.

Magat, W.A., and W.K. Viscusi. 1993. Informational Approaches to Regulation. Cambridge, Mass: MIT Press.

Magat, W.A., A.J. Krupnick, and W. W. Harrington. 1986. Rules in the Making: A Statistical Analysis of Regulatory Agency Behavior, Washington D.C: Resources for the Future.

Mahe, L.P. 1997. Environment and quality standards in the WTO. New protectionism in agricultural trade. European Review of Agricultural Economics 24(3–4):480–503.

Mahe, L.P., and F. Ortalo-Magne. 1998. International co-operation in the regulation of food quality and safety attributes. Paper presented at the OECD Workshop on Emerging Trade Issues in Agriculture, Paris, Oct. 25–27.

Mazurek, J.V. 1996. The Role of Health Risk Assessment and Cost-Benefit Analysis in Environmental Decision Making in Selected Countries: An Initial Survey. Discussion Paper 96-36. Washington D.C.: Resources for the Future.

OECD (Organization for Economic Cooperation and Development). 1997a. Regulatory Reform in the Agro-Food Sector. Pp. 233–274 in Regulatory Reform Volume I: Sectoral Studies. Paris: OECD.

OECD (Organization for Economic Cooperation and Development). 1997b. The Costs and Benefits of Food Safety Regulations: Fresh Meat Hygiene Standards in the United Kingdom. Paris: OECD.

OECD (Organization for Economic Cooperation and Development). 1999. Food safety and quality issues: trade considerations. Paris: OECD.

Olson, E. 1998. Critics Say World Trade Group Disregards Environment. The New York Times, May 16.

Orden, D., and D. Roberts, eds. 1997. Understanding Technical Barriers to Agricultural Trade. Proceedings of the International Agricultural Trade Research Consortium, St. Paul: University of Minnesota.

Pollak, R.A. 1995. Regulating risks. Journal of Economic Literature, 33(1):179–191.

Pollak, R.A. 1998. Imagined risks and cost-benefit analysis. American Economic Review, Papers and Proceedings 88(2):376–379.

Powell, M. 1997. Science in Sanitary and Phytosanitary Dispute Resolution. Discussion paper 97-50. Washington D.C.: Resources for the Future.

Public Citizen Global Trade Watch. 1998. Comments of Public Citizen Inc. regarding U.S. preparations for the World Trade Organization's ministerial meeting, fourth quarter 1999. Public Citizen Inc., Washington, D.C., Oct. 22.

Rege, V. 1994. GATT law and environment: related issues affecting the trade of developing countries. Journal of World Trade 28(3):95–169.

Roberts, D. 1997. Implementation of the WTO Agreement on the application of sanitary and phytosanitary measures. Paper presented at the International Agricultural Trade Research Consortium Meeting, San Diego, Calif., Dec. 14–16.

Roberts, D., and K. Deremer. 1997. Overview of Foreign Technical Barriers to U.S. Agricultural Exports. Commercial Agriculture Division, Economic Research Service Staff paper AGES-9705. Washington D.C.: U.S. Department of Agriculture.

Salter, L. 1988. Mandated Science: Science and Scientists in the Making of Standards, Dordecht, The Netherlands: Kluwer.

Silverglade, B. 1998. Should the SPS Agreement be amended? A modest proposal to restore public support. A paper presented at the Ceres Conference on Politicizing Science: What Price Public Policy? Georgetown University Public Policy Institute, Washington, D.C., April 4.

Stiglitz, Joseph. 2000. Economic of the public sector. 3rd Ed. W.W. New York: Norton and Co.

Thilmany, D.D., and C.B. Barrett. 1997. Regulatory Barriers in an Integrating World Food Market. Review of Agricultural Economics 19(1):91–107.

Thornsbury, S., D. Roberts, K. Deremer, and D. Orden. 1997. A first step in understanding technical barriers to agricultural trade. Paper presented at the XIII International Conference of Agricultural Economists, Aug.10–16. Sacramento, Calif.

Viscusi, W.K. 1993. The Value of Risks to Life and Health. Journal of Economic Literature 31: 1912–1946.

Viscusi, W.K., J.M. Vernon, and J. E. Harrington, Jr. 1995. Economics of Regulation and Antitrust. Cambridge, Mass.: Massachusetts Institute of Technology Press.

Vogel, D. 1995. Trading up. Consumer and Environmental Regulation in a Global Economy. Cambridge, Mass.: Harvard University Press.

Part III

Case Studies

8

Case Study 1: Meat Slaughtering and Processing Practices

THE DANISH APPROACH TO FOOD SAFETY ISSUES RELATED TO PORK PRODUCTS

BENT NIELSEN
Section for Zoonotic Diseases, Veterinary and Food Advisory Service,
Federation of Danish Pig Producers and Slaughterhouses,
Copenhagen, Denmark

DANISH CONSUMERS' PERSPECTIVES ON FOOD SAFETY

Danish consumers focus increasingly on food safety and the welfare of all types of livestock. The consumers' attitude to food safety and animal welfare is based on both correct and factual information, and also to a very large extent on beliefs and old wives' tales. Both the media and Danish politicians have long since found out that food safety and animal welfare are issues that sell. Hardly a month goes by without a newspaper publishing a critical story about the poor quality of food products, the use of antibiotics or the minimal space allocated to livestock in modern livestock buildings.

Over the past 20 years, Danes have become increasingly distanced from livestock production in Denmark, and the rural population now accounts for less than 5 percent of the Danish population. With almost no direct experience of livestock production, most consumers are at the mercy of the information or misinformation provided by the media and the politicians. Below follows an outline of the most important factors affecting Danish consumers' perception of food safety.

Comparison with Sweden

Denmark is a neighbor of Sweden. Danish politicians and the Danish media often compare Danish and Swedish conditions because both countries are very similar in a number of respects.

It is, therefore, natural for Danish consumers and their professional and industrial bodies to look at food safety conditions in Sweden and to make comparisons with conditions in Denmark. Over the past 30 years, Sweden has fought a very active battle against *Salmonella* in its livestock production. The occurrence of *Salmonella* in Swedish livestock production is close to nil, and, in reality, Swedish food products are considered to be *Salmonella* free. The use of antibiotic growth promoters has been banned since 1986, and the use of antibiotics for therapeutic purposes is heavily restricted. Finally, a number of special welfare requirements have been introduced in respect to livestock production in recent years, such as a minimum floor area per animal, which is approximately 30 percent larger than the Danish equivalent.

CONSUMER REQUIREMENTS OF DANISH MEAT

Danish consumers make certain requirements and have certain expectations of the meat they buy. The most important requirements are the following (as in Sweden):

- an absence of zoonotic agents such as *Salmonella;*
- an absence of chemical residues such as antibiotics, hormones, pesticides, etc.; and
- a wish for good animal welfare throughout the life of the pig, including slaughtering.

Zoonotic Agents

Diseases that can be transmitted from animals to humans and vice versa are called zoonoses. Danes are very sensitive to the occurrence of zoonoses both in meat produced in Denmark and in imported meat. It is the general attitude among consumers that zoonotic agents must not be found in Danish food products.All Danish consumers know of *Salmonella*, and they know that the bacteria do occur in meat from time to time. In the past year, there has been considerable focus on the difference between the levels of *Salmonella* in Danish and imported meat. Danish meat has a very low prevalence of *Salmonella* in comparison with other countries, with the exception of the other Scandinavian countries. This has resulted in a demand for the testing of imported meat.

This trend was reinforced considerably in 1998, following increased focus on the multiresistant *Salmonella typhimurium* DT-104. DT-104 is typically resistant to ampicillin, chloramphenicol, streptomycin, sulfonamides and tetracycline, but has a high ability to develop further resistance to quinolones and trimethoprim (Danish Zoonosis Centre, 1999). It can, therefore, be more difficult to treat humans who develop DT-104 salmonellosis with antibiotics. In Europe, the typical first-choice drug used to treat serious salmonellosis in humans is a quinolon, ciprofloxacin. In cases when the salmonellosis has been caused by a quinolon-resistant DT-104 strain, there is a risk of treatment failure, which may be extremely critical for the patient. The multiresistant DT-104 is frequently found in most European countries (with the exception of Scandinavia) the Far East, and increasingly also the United States .

DT-104 is extremely rare in Danish livestock production. Between 1991 and October 1998 only 25 swine herds had been affected in all of Denmark, and the occurrence of DT-104 in pork is correspondingly low, approximately 1 out of 35,000 meat samples examined.

The demand from Danish politicians and thereby also Danish consumers is for the complete absence of multiresistant DT-104 bacteria in all meat, Danish as well as imported. Some Danish pig herds are being stamped out in order to eradicate DT-104 on the farms. All meat from a DT-104 herd must be subjected to heat treatment to prevent exposure of consumers to DT-104 contamination. This represents a dramatic sharpening of the population's view on *Salmonella*.

The general attitude to the presence of other *Salmonella* bacteria in food products has been sharpened similarly in the course of 1998. The level of *Salmonella* in Danish pork is very low. Over the past four years, *Salmonella* has only been found in approximately 1 percent of the 28,000 samples examined annually by the Danish slaughterhouses. However, Danish consumers still believe that this is too high, although most other industrialized countries have *Salmonella* in 5–30 percent of their pork.

Danish consumers have also heard of the *Campylobacter* bacteria. However, this zoonosis is primarily linked with poultry, not pork. Very few consumers know of other zoonoses in food products.

When discussing the issue of zoonotic diseases, it is worth mentioning that Denmark demands a non-discriminating testing of imported meat. Five percent of pork and beef batches, and 10 percent of poultry batches are tested for the presence of bacteria using cultural methods. Testing of imported meat is performed on the same level as testing of domestic meat which, consequently, is designated as a non-discriminating testing. Test results for both domestic and imported meat are published on the homepage of the Danish Department of Agriculture.

Residues of Antibiotics, Hormones, Pesticides, Heavy Metals, and Others

During the past three years there has been considerable focus on the use of antibiotics in livestock in Denmark, especially antibiotic growth promoters. Consumers are concerned about two aspects: the risk of residues and the development of antibiotic-resistant bacteria transmissible to humans.

Many consumers share the misunderstanding that the use of antibiotic growth promoters leads to residues in meat. The Danish Veterinary Services examine approximately 20,000 carcasses of swine annually for a large number of residues of antibiotics. Only 3–5 carcasses are found to be positive. Thus, antibiotic residues in Danish pork are not a real problem.

There has been an increasing fear among Danish and Swedish microbiologists that the long-term use of antibiotic growth promoters can promote the development of antibiotic-resistant bacteria transmissible to humans. Researchers fear that, in the long term, this may reduce the possibilities for treating infections in humans (Bager and Emborg, 1999). Consequently, use of the antibiotic growth promoter avoparcin has been banned since 1996. Since March 1, 1998, Danish pig producers have introduced a voluntary ban on the use of growth promoters in pigs weighing more than 35 kg. At the same time, Danish cattle and poultry producers introduced a complete ban on the use of growth promoters. The Danish government wishes to stop the use of growth promoters as soon as possible and on September 1, 1998, introduced a special tax on antibiotic growth promoters. This means that it no longer makes sense economically to use them.

In September 1998, the European Union (EU) conference "The Microbiological Threat" was held in Copenhagen. The object of the conference, which received much attention in the media, was to harmonize the policies of the EU countries in the area of antibiotics. As a result of the conference, Danish pig producers decided to stop the use of antibiotic growth promoters completely within the next year. At the moment, there has already been a considerable decrease in the number of farms using antibiotic growth promoters for piglets. This is considered a big step in the right direction by consumers, even though there is no conclusive documentation on the risk of continued use of antibiotic growth promoters for swine. As in the case of the zoonotic agents, the consumers' attitude is that consumers should be given the benefit of the doubt.

Consumers are not particularly concerned with residues of hormones, pesticides, and heavy metals as pork has not been linked with these issues by the Danish media. The Danish Veterinary Services regularly examine carcasses of swine for these residues, but so far there have been no positive findings.

WELFARE

For Danish consumers, animal welfare and food safety are closely related. The demand for organic vegetables, grain products, eggs, milk, cheese, and meat has increased dramatically. An increasing proportion of Danish consumers is convinced that a high level of animal welfare equates to an absence of zoonotic agents, antibiotic residues, etc. There is no doubt that increased animal welfare can leave consumers with a better moral taste in their mouths. However, there is no link between a high level of animal welfare and the absence of zoonotic agents. So far, studies in Denmark have shown that the levels of *Salmonella* in conventional pigs and special welfare pigs are the same. This is a message that consumers find hard to accept.

CONCLUSIONS

Danish and other Scandinavian consumers' demands for food safety are increasing. The ideal situation would be an absence of zoonotic agents and all types of chemical residues. At the same time, the animal must have been reared under optimum welfare conditions. In this way, it should be possible to ensure that the food products bought by consumers are very safe and also that consumers can eat the meat from the animal with a clear conscience.

For the most part, Danish pig producers have been able to comply with the wishes of Danish consumers. Seen in an international perspective, the level of zoonotic agents is very low, chemical residues are virtually nonexistent, and pig producers are in the process of adopting more welfare-friendly systems for their livestock buildings.

AN UPDATE ON THE DANISH *Salmonella* REDUCTION PROGRAM

In 1993 a preliminary *Salmonella* surveillance program of slaughter pig herds was initiated with a permanent program being established in January 1995. In 1998, the program was revised and new initiatives have been implemented.

The aim of the compulsory program is to reduce the prevalence of *Salmonella* in slaughter pig herds and pork. Here I describe the monitoring program of slaughter pig herds, the regulations that the herd owners are required to follow, and the results achieved so far. An update of the system has been presented recently at the International Pig Veterinary Society (IPVS) Conference, Birmingham, United Kingdom (Emborg et al., 1998).

Materials and Methods

The *Salmonella* reduction program consists of the following parts:

(1) Serological monitoring of all herds producing more than 100 slaughter pigs per year.
(2) Assignment of herds into one of three levels (1, 2, or 3) based on the prevalence of seroreactors.
(3) Mandatory advising and elaboration of a *Salmonella* intervention plan for all herds in levels 2 and 3.

Furthermore, the program includes monitoring of *Salmonella* in animal feed, breeding and multiplying herds, and the prevalence in pork products. Finally, pigs from level 3 herds are slaughtered under special hygiene precautions (none of these parts of the program are described in this paper).

The above-mentioned parts of the program are performed as follows:

(1) The herd-monitoring scheme makes use of an indirect enzyme-linked immunosorbent assay (ELISA) based on a combination of the lipopolysaccharide (LPS) antigens O:1, 4, 5, 6, 7, and 12 (the so-called mix-ELISA). The assay was developed by the Danish Veterinary Laboratory for the use on serum, but has been modified to be used also on meat juice (Nielsen et al., 1995). Meat juice is obtained when frozen meat samples from slaughter pigs are thawed. The slaughterhouses collect meat samples continuously from about 16,000 herds with samples being taken at random from each herd. Each quarter, between 8 and 60 meat samples are collected from each herd, with the number of samples determined by the number of pigs delivered for slaughter. Around 800,000 meat juice samples are examined annually for *Salmonella* antibodies at the laboratory (Mousing et al., 1997).

(2) The result of the examination of the meat juice samples is summarized monthly for the individual herd. Based on the proportions of seroreactors during the previous three months, the herds are assigned to one of three levels. Level 1 herds have no or very few seroreactors, level 2 herds have a relatively high proportion of seroreactors, whereas level 3 herds have an unacceptably high proportion of seroreactors (Mousing et al., 1997). Both the herd owners and the slaughterhouses are informed monthly about the *Salmonella* level of the herds. When a herd is placed in levels 2 or 3, the herd owner must initiate a *Salmonella* intervention plan (Mousing et al., 1997).

(3) Since January 1995, owners of the herds assigned to levels 2 and 3 are requested by the slaughterhouse to seek advice on how to reduce the prevalence of *Salmonella* in the herds. The herd owner, a veterinary surgeon, and a pig consultant must elaborate a herd-specific intervention plan, otherwise the slaughterhouse will collect a penalty per slaughtered pig delivered until the plan has been elaborated and received by the slaughterhouse. Three months after the assignment to levels 2 and 3, the veterinary surgeon and the pig consultant must certify that the program agreed upon is being followed. If not, the slaughterhouse will again collect a penalty per slaughtered pig (4 percent of the value of each finisher slaughtered). If the herd remains in levels 2 or 3, or the herd is reassigned to levels 2 or 3 six months after the first assignment, it is required that the owner again seek advice on how to reduce the *Salmonella* prevalence in the herd as described above.

From August 1996, the requirements of the intervention in the levels 2 and 3 herds were increased. Ordered by the Danish Veterinary Services, these requirements include that a sufficient number of pen fecal samples must be collected and analyzed in order to clarify the distribution of *Salmonella* in the herd. Based on these results, an appropriate intervention plan must be prepared (Emborg et al., 1997).

In addition, the slaughterhouses announced in July 1996 that from January 1997, a slaughtering fee would be charged on all herds assigned constantly to level 3 for more than six months. The fee will be collected until the herd is assigned to levels 1 or 2 (Emborg et al., 1997).

Results and Discussion

The results of the Danish *Salmonella* surveillance program comprising approximately 16,000 slaughtered pig herds from June 1995 to August 1998 are presented below.

Seropositive Meat Juice Samples

The number of seropositive meat juice samples varied between 4 and 7 percent in the period 1995 to the end of 1997, while a significant decrease was observed from October 1997 to June 1998, reaching a minimum at 2.3 percent (which is considered a very low prevalence). Since June 1998, the number of seropositive meat juice samples has remained below 3 percent. It is assumed that the observed decrease in seropositive meat juice samples is a consequence of more effective *Salmonella* reduction strategies at the farm level.

Level 2 and 3 Herds

Throughout the surveillance period the percentages of levels 2 and 3 herds ranged from 2.4 to 4.3 percent and 1.1 to 2.3 percent, respectively. Although the percentages of herds assigned to level 2 varied, the percentage did not decrease significantly before the spring of 1998, as a result of the decreasing number of seropositive meat juice samples. The number of level 2 herds has remained below 3 percent in the period February to September 1998.

From August 1996 to March 1997, a significant decrease was found in the proportion of level 3 herds ($\beta = -0.032$, $P = 0.011$) (Emborg et al., 1997). However, since March 1997 no further decrease occurred, with the proportion of level 3 herds remaining between 1.2 and 1.8 percent. Surprisingly, the significant decrease in meat juice samples in 1998 did not decrease the number of level 3 herds.

The decrease in the proportion of level 3 herds from August 1996 to March 1997 may be associated with the obligatory requirements to collect and analyze pen fecal samples for *Salmonella*, which were introduced in August 1996, and the announcement in July 1996 that a slaughtering fee would be effective starting in January 1997. It appears that a further decrease in the proportion of level 3 herds and an additional decrease in the proportion of level 2 herds are possible only if the number of chronically infected herds is reduced.

During the surveillance period, owners of 3,955 herds (about 25 percent of the 16,000 herds) have been requested to seek advice on how to reduce the *Salmonella* prevalence in the herds (Table 8-1). In 1,747 (44 percent) of the herds, the high prevalence of *Salmonella* did last more than six months and the consequences were two or more requirements to seek advice. In 233 (5.9 percent) herds the problems with *Salmonella* have been so persistent that 5 to 7 requirements to seek advice have been necessary.

TABLE 8-1. Monitoring Results of the *Salmonella* Reduction Program

No. of Times a Herd Owner was Required to Seek Advice on Reducing the *Salmonella* Prevalence in the Slaughter Pig Herds	No. of Slaughter Pig Herds	Percent of Total No. of Slaughter Pig Herds
1	2,208	55.8
2	823	20.8
3	436	11.0
4	255	6.4
5	141	3.6
6	85	2.1
7	7	0.2
Total	3,955	100

New Initiatives

The number of herds with more than five requirements is clearly unsatisfactory. Too many of the chronically infected herds stay too long in level 3, and the finishers (pigs that reached the slaughter weight) consequently have to be slaughtered under increased hygiene precautions. In spring 1998, the Federation of Danish Pig Producers and Slaughterhouses decided to increase the pressure on the chronically infected herds and level 3 herds in general to reduce the number of finishers for special hygiene slaughter. Two new initiatives were introduced by September 1, 1998: second-opinion advisers and a level 3 slaughter fee.

Second-Opinion Team

If the fifth requirement is given within 36 months, two second-opinion advisers must participate in the preparation of the intervention plan. The second-opinion advisers consist of a team of five veterinarians and five swine consultants who are specialists on *Salmonella*. The cost of a veterinary advice amounts to approximately $200–300 (U.S.).

Level 3 Slaughter Fee

In addition, the slaughterhouse will collect a fee per level 3 finisher slaughtered under special hygiene conditions. The reason for collecting the slaughter fee is the extra spending due to slaughter under special hygiene precautions. The estimated cost per finisher for a special slaughter is $25. From September 1998, the slaughter fee will be calculated as presented in Table 8-2.

TABLE 8-2. Special Slaughter Fees per Finisher

Months in Level 3	Fee per Finisher ($)
0–3	0
4–6	3.30
7+	6.30

To return to $0 per level 3 finisher, the herd must not be assigned to level 3 during the next 12 months. For example, if a herd stays five months in level 3, the farmer will be deducted $3.30 for every finisher in months 4 and 5. If the herd subsequently is assigned to level 1 for the next three months and then goes back to level 3 for an additional four months, the farmer will be deducted $3.30 per finisher for the sixth month (month 9 of the year) in level 3 and $6.60 per finisher for the seventh to ninth months (months 10–12 of the year) in level 3.

Future goals for the Danish *Salmonella* reduction program is to reach a level of less than 0.5 percent of *Salmonella* in pork by the year 2001. This will be achieved by intensified control pre-harvest and increasing hygiene on slaughter plants.

REFERENCES

Bager, F. and Emborg, H.-D. 1999. Danish Integrated Antimicrobial Resistance Monitoring and Research Programme (DANMAP) 98 Report: Consumption of antimicrobial agents and occurrence of antimicrobial resistance in bacteria from food animals, food and humans in Denmark. Copenhagen, Denmark: Danish Zoonosis Centre.

Danish Zoonosis Centre. 1999. Annual report on zoonosis in Denmark 1998. Hald, T., H.C. Wegener, and B.B. Jørgensen, eds. Copenhagen, Denmark: Danish Zoonosis Centre.

Emborg H.-D., A. C. Nielsen, P. Thode Jensen, J. P. Nielsen, and J. Mousing. 1997. The Danish *Salmonella* Surveillance Programme of Slaughter pig herds. In the Proceedings of the 8th International Society of Veterinary Epidemiology and Economics (ISVEE), 07.14.1–3. Paris.

Emborg, H.-D., V. Møgelmose, and B. Nielsen. 1998. Status of the Danish *Salmonella* Surveillance Programme of Slaughter Pig Herds. Birmingham, UK: International Pig Veterinary Society (IPVS).

Mousing J., P. Thode Jensen, C. Halgaard, F. Bager, N. Feld, B. Nielsen, J.P. Nielsen, and S. Bech-Nielsen. 1997. Nation-wide *Salmonella enterica* surveillance and control in Danish slaughter swine herds. Preventive Veterinary Medicine 29:247–261.

Nielsen B., D.L. Baggesen, F. Bager, J. Haugegaard, and P. Lind. 1995. The serological response to *Salmonella* serovars typhimurium and infantis in experimentally infected pigs. The time course followed with an indirect anti-LPS ELISA and bacteriological examinations. Veterinary Microbiology 47:05–218.

INTERNATIONAL HARMONIZATION UNDER THE SPS AGREEMENT

BRUCE A. SILVERGLADE
Center for Science in the Public Interest, Washington, D.C.

I would like to thank the sponsors of today's conference for the opportunity to speak to you regarding how consumer organizations in the United States view the international harmonization of food safety regulations.

The Center for Science in the Public Interest (CSPI) is a consumer advocacy organization based in Washington, D.C. The center is supported by almost one million subscribers to its magazine, *Nutrition Action Health Letter*, which reports on food safety and nutrition issues. CSPI was formed in 1971 and over the past two decades has campaigned for the elimination of hazardous food additives such as sulfiting agents, worked to improve meat and poultry inspection, and fought for mandatory nutrition labeling requirements.

To further its role in international issues, CSPI became a recognized observer at the *Codex Alimentarius* Commission and formed a new organization called the International Association of Consumer Food Organizations (IACFO), which is an international coalition of consumer groups that work primarily on food safety and nutrition issues. Charter members include the Food Commission U.K., based in London, and the Japan Offspring Fund, based in Tokyo. IACFO has filed comments with various governments on the labeling of genetically engineered foods (IAFCO, 1998), issued a report on the regulation and marketing of functional foods (IAFCO, 1999), and has participated in *Codex* committee meetings.

Today, I wish to present the consumer viewpoint on the process of international harmonization under the Sanitary and Phytosanitary (SPS) Agreement (see Appendix A). I will address whether we are harmonizing in an upward or a downward direction. To illustrate the concerns of consumers, I will review some of the recent activities of the *Codex* Alimentarius Commission, which is officially recognized under the SPS agreement as a source of international standards that can be used by the World Trade Organization (WTO) to resolve trade disputes. I will also examine how equivalency agreements developed pursuant to the SPS Agreement can affect the international harmonization process. Lastly, I will briefly discuss the role that science and other factors play in SPS decisions and draw some conclusions from CSPI's experience to date.

It was actually after Congress passed mandatory nutrition labeling legislation in 1990 (Pub. L. No. 101-535) that CSPI began to open its eyes to international issues. After the nutrition labeling legislation took effect, the European Union (EU) began to complain that the new law was a trade barrier (European Commission, 1997). In response, CSPI began looking into international trade issues to see if some of the biggest consumer victories in the United States could become the victim of trade disputes.

My remarks today, however, should not be construed in any manner as an attack on free trade. Free trade promotes an efficient allocation of resources. It increases the variety of goods available to consumers and it lowers the prices of those goods. And free trade certainly encourages better world citizenship by facilitating peaceful cooperation and exchange. We are in a global economy to stay, and international harmonization of regulatory requirements is necessary. But are we harmonizing upward or downward? In what direction are we going?

To the extent that international harmonization elevates health and safety regulations to a consistent level of excellence, consumers worldwide are well served. Under this scenario, international standards would incorporate the best features of national standards that consumer organizations believe provide the public with the highest levels of protection. However, if harmonization tends to reduce standards to some acceptable international norm, then consumer health and safety may be jeopardized regardless of the economic benefits brought about by free trade. President Clinton has recognized this, and in a 1998 speech last summer to the World Trade Organization he has called for a "leveling up," in his words, of consumer protection regulations and not a leveling down (Office of the President, 1998).

But what is actually happening? We believe that international harmonization is leading to a leveling down of consumer standards. This is occurring for several reasons.

First, under the SPS Agreement, international standards serve as a ceiling, not a floor. They are a maximum rather than a minimum. And there is nothing in the SPS Agreement that requires the setting of a minimum floor that countries can exceed; it is just the opposite. So there is implicit pressure for downward harmonization built into the SPS Agreement.

Second, the SPS Agreement was adopted to facilitate trade, not to raise health and safety standards. The SPS is not a public health agreement, it is a business-oriented trade agreement that is supposed to reduce regulation and make it easier for companies to trade internationally. And, in fact, many of the processes involved with the SPS Agreement, particularly the proceedings of the *Codex Alimentarius* Commission, have become forums for deregulation.

Third, public participation by consumer groups, environmental groups, and others in international proceedings is limited for obvious reasons related to resources and logistics. We hope that this will change, but at the present time, lack of consumer input is certainly one of the factors that we believe is leading to downward harmonization.

I would now like to provide some illustrations of where downward harmonization is occurring. First, *Codex* has finalized a standard that does not require pasteurization of cheese (FAO/WTO, 1999). Pasteurization has been a hallmark of food safety in the United States. However, the *Codex* standard is based on practices common within the EU. This standard represents an example of where the U.S. has been forced to accept an international standard that fails to provide the same level of public health protection afforded by domestic regulatory requirements. While U.S. government officials are quick to point out that nothing in the SPS Agreement requires us to adopt the *Codex* standard, they fail to note that other countries have a right of action to challenge current U.S. regulatory requirements in this area as a trade barrier, and that the U.S. would be hard pressed to provide a scientific justification for such requirements in light of the *Codex* standard.

Second, as I mentioned, mandatory nutrition labeling in the United States has been attacked as a trade barrier. Proponents of this view frequently cite the *Codex* guidelines for nutrition labeling (FAO/WTO, 1993), which only require disclosures of such information if the manufacturer makes a nutrition claim. Because the U.S. requirement exceeds the *Codex* guidelines there have been proposals that the United States permit imports of foods that have some other, lesser form of nutrition information on the label as opposed to the full list of nutrients mandated by Congress. Label disclosures must be standardized to be effective. Public health and consumer organizations cannot teach people to use nutrition labels if they are not presented in a consistent format. Allowing imported products with some other country's nutrition label would jeopardize the objectives of the U.S. law and represent another example of downward harmonization.

Third, *Codex* has adopted standards, opposed by the United States, for natural mineral water (FAO/WTC, 1997a) that allow greater levels of contaminants than permitted under U.S. Food and Drug Administration (FDA) regulations (21 CFR 165.110). The adoption of the *Codex* standard was quite a loss for the FDA, which fought for years to set stringent bottled water standards in the United States and now may be confronted with demands to permit the import of products that fall below those standards.

Certainly there is a potential that international harmonization can raise standards. As David Vogel points out, we can trade up (Vogel, 1995); but whether this is what is happening is questionable.

Two additional examples illustrate our concerns. The first is the WTO's decision in the growth hormones case (WTO, 1998) and the second is equivalency agreements developed by the U.S. Department of Agriculture (USDA) regarding meat and poultry inspection. The downward harmonization problem is illustrated by both of these matters.

The hormone decision is obviously a very important decision under the SPS Agreement. CSPI has not campaigned against the use of hormones in the United States. We recognize that there is a significant percentage of Americans who want to buy organic or natural beef and dislike hormones as much as the Europeans, but in general, we have not made an issue about the use of hormones. For the sake of this discussion, I will assume that there is no human health risk posed by hormone use in the United States.

Nevertheless, the EU does not want to buy U.S. beef, and this is not simply a protectionist issue. Certainly there is a degree of trade protectionism in the EU position, but that position is supported by the European public for other reasons. It would, in fact, be difficult for the EU to maintain such a protectionist stance if it was not the subject of popular support. Popular support for the hormone ban in Europe can be traced to various historical experiences and cultural values. There were problems in Italy with the use of hormones. The European consumer remembers that disaster, and distrusts government authorities even more in the wake of the mad cow disease fiasco (Echols, 1998). And in regard to cultural values, there is the view that America, through its agricultural exports, is trying to McDonaldize the EU food supply.

Therefore, based on different historical experiences and cultural values, most European consumers have come to oppose the use of hormones in cattle. The point here is that the distaste for hormone-treated beef in the EU is real. Simply dismissing such attitudes as trade protectionism, as many U.S. officials do, is not useful because it does not address the views of European consumers. Moreover, simply trying to force the issue at the WTO is counterproductive and threatens to destabilize the entire world trading system.

CSPI is concerned about the hormone decision from another standpoint. The decision could be interpreted as limiting the right of a nation to establish a zero-risk standard under certain circumstances. For example, the United States maintains the Delaney Clause to the Food, Drug and Cosmetic Act, which sets a zero-risk standard for cancer causing food and color additives (21 USC 348(c)(3)(A), 379e(b)(5)(B)). This regulatory approach embodies the philosophy that in some cases, the benefits from permitting a substance in our food supply (such as a artificial color additive that is used primarily in non-nutritious "junk" food) may be so minimal that no risk of cancer, even an extremely small one, can be justified. Consumer groups support the Delaney Clause, not because it is science based, but because it is based on this important principle and represents an insurance policy against weak regulation in times of budget crunches or competing priorities. It forces the FDA to make tough policy decisions when the agency might be pressured by industry to look the other way and ignore certain risks. But we would be hard pressed to argue before the WTO that it is science based.

The WTO's decision in the hormone case could have a boomerang effect and come back to haunt the United States by jeopardizing consumer protection requirements like the Delaney Clause. The SPS Agreement was essentially written as a business document to increase agricultural exports, not to protect public health. Insufficient thought was put into it at the time it was drafted as to how it might hurt us in the United States in certain areas that we believe are important.

Another example of where the principle of a downward harmonization seems to be at work involves equivalency agreements. The *Codex Alimentarius* Commission approved, over the objections of the U.S. government, equivalency guidelines for the establishment of import and export inspection and certification systems (FAO/WTO, 1997b) that do not require the use of government employees to inspect food products. The United States has long relied on government employees to inspect meat and poultry. The *Codex* guidelines approved in 1997 can be interpreted as sanctioning the use of company employees to inspect such products. It was pushed very heavily by

Australia, which has a conservative government in power, is deregulating quite actively, and favors the use of company employees as opposed to government employees to conduct inspections. CSPI feared that the *Codex* guidelines that were adopted would be used to pressure the United States to accept imports from countries that rely on company employees for inspection responsibilities.

In fact, that is what has precisely happened. USDA, citing the *Codex* guidelines, has finalized equivalency agreements with numerous other countries that do not mandate the use of government paid inspectors. This action is inconsistent with domestic regulatory policy. For example, USDA's own regulations require *Salmonella* testing by government employees (7 CFR 381.94(b); 9 CFR 310.25(b)). The rules took effect in January 1998 for producers that have 500 or more employees. Fifteen countries that export meat to the United States have factories that employ 500 or more employees (House of Representatives, 1999). Under the equivalency agreements negotiated by USDA, these nations will agree to perform the *Salmonella* testing but, in many cases, will permit firms to use company employees to perform the necessary tests. In explaining their decision, USDA officials cited the *Codex* standard as a factor in their thinking and stated very plainly that, under the equivalency provisions of the SPS Agreement, they would not require other countries to use government employees to do the *Salmonella* testing. This decision establishes a double standard that not only consumer groups, but U.S. producers as well, should be concerned about.

It does not have to be that way. Through equivalency agreements, we can certainly learn how to improve food safety requirements. The United States does not necessarily have the strongest requirements in every area. The U.S. food industry likes to say that it has the safest food supply in the world, but that is no longer true across the board. Therefore, we can certainly benefit from equivalency agreements if they are used to raise consumer protection standards. Unfortunately, they are currently being used to merely facilitate trade at the cost of lowering consumer protection requirements. It is an issue that concerns CSPI very much, and I can assure you that we will be working a great deal on it.

Some will argue that harmonizing upward is too costly, especially for developing countries. The SPS agreement specifies that developing countries should receive technical assistance in order to comply with their SPS obligations. In reality, such technical assistance has rarely been provided. This must change. In order to maintain public support for international harmonization among consumers in developed countries, such nations must be required to provide technical assistance to developing countries so as to enable them to comply with world class standards.

Finally, I will briefly address the role of science in policy making under the SPS Agreement. Obviously, science has to take a leading role, but science has its limits when it comes to risk management decisions and it is not value free. Risk assessments are based on assumptions and we have heard that these can be rooted in cultural values. Just the decision to do a risk assessment on a particular substance, but not on another substance, is a subjective judgment that may be based on cultural values. And although science has to play the leading role in informing policy decisions, other factors certainly enter into the equation. This happens every day at the *Codex Alimentarius* Commission, which plays a major role under the SPS Agreement. How else can one explain such close votes at *Codex* approving the use of growth hormones in cattle? If *Codex* were proceeding strictly on the basis of scientific consensus, such matters, by definition, would not be the subject of close votes. Voting would be unnecessary. But *Codex* decisions are not simply based on scientific consensus; factors other than science are most certainly entering into the decision-making process.

One of these factors, so obvious that perhaps we do not see it, is trade concerns. We can debate the extent to which consumer and environmental concerns should be considered along with scientific factors, but trade concerns are already being taken into account in what are purportedly purely scientific decisions. If we are considering trade concerns at *Codex*, then we should certainly be considering consumer concerns, some of which are based on cultural differences, as well.

Cultural differences play a very key role in explaining what some may think are just protectionist attitudes or what some may say is the misuse of science by the press or politicians. To many of the economists participating in today's conference, cross-cultural disputes simply look like trade protectionism. Every time an SPS dispute arises, they say, "Well, it's just disguised protectionism." It is actually much more complicated. To many of the scientists participating in today's conference, these disputes may be attributed to consumer activists trying to generate publicity or politicians trying to garner votes. Those parties certainly play a key role, but SPS disputes cannot simply be chalked up to protectionist attempts to grab media attention, or to political pressures.

Lawyers may say that SPS disputes are essentially legal disputes. The SPS Agreement is an international law, but SPS disputes do not only involve controversies over the meaning of legal terms. Cultural differences play a very, very large role. And there is no doubt that culture will continue to play a key role in what some may regard as purely scientific or economic issues.

In conclusion, let me say that international harmonization can be a positive experience. It can lead to the adoption of international standards that embody the best consumer protection policies from around the world. The potential is there. As we proceed with the global economy, we really have only one option: to harmonize upward. Consumers will see any other course of action as untenable. The challenge is there and how we meet it will not only affect the future of food regulation, but whether public support for the world trading system will grow or diminish even further than it has already.

REFERENCES

Echols, M.A. 1998. Food Safety Regulation in the European Union and The United States: Different Cultures, Different Laws. Columbia Journal of European Law 4:525–543.

European Commission. 1997. Report on United States Barriers to Trade and Investment. Brussels, Belgium: European Commission.

FAO/WHO (Food and Agriculture Organization of the United Nations/World Health Organization). 1993. *Codex Alimentarius* Commission: *Codex* Guidelines on Nutrition Labelling, CAC/GL 2-1985 (Rev. 1-1993). Joint FAO/WHO Food Standards Programme, Rome: FAO.

FAO/WHO (Food and Agriculture Organization of the United Nations/World Health Organization). 1997a. *Codex* Standard for Natural Mineral Waters, ALINORM 97/20, Appendix II. Joint FAO/WHO Food Standards Programme. Rome: FAO.

FAO/WHO (Food and Agriculture Organization of the United Nations/World Health Organization). 1997b. Guidelines for the Development of Equivalence Agreements Regarding Food Import and Export Inspection and Certification Systems, CAC/GL 34-1999. Joint FAO/WHO Food Standards Programme. Rome: FAO.

FAO/WHO 1999. *Codex* Alimentarius Commission: *Codex* General Standard for Cheese, *Codex* Standard A-6-1978 (Rev. 1-1999). Joint FAO/WHO Food Standards Programme. Rome: FAO.

IACFO (International Association of Consumer Food Organizations). 1998. Comments to Ministry of Agriculture, Forestry, and Fisheries, Japan. October 9. Washington, D.C.: IACFO.

IACFO (International Association of Consumer Food Organizations). 1999. Functional Foods: Public Health Boon or 21st Century Quackery? Washington, D.C.: IACFO.

The Office of the Federal Register, National Archives and Records Administration, (OFR/NARA), 1999. 7 CFR 381.94(b); 9 CFR 310.25(b); 21 CFR 165.110. Washington, D.C.: U.S. Government Printing Office.

Office of the President. 1998. Speech: Remarks by the President at the Commemoration of the 50TH Anniversary of the World Trade Organization. May 18. Washington, D.C.: The White House.

U.S. House of Representatives. 1999. Hearings Before a Subcommittee of the Committee of Appropriations, Agriculture, Rural Development, Food and Drug Administration, and Related Agencies Appropriations for 2000. Washington, D.C.: U.S. Government Printing Office.

Vogel, D. 1995. Trading Up: Consumer and Environmental Regulation in a Global Economy. Cambridge, Mass.: Harvard University Press.

WTO (World Trade Organization). 1997. WT/DS26/AB/R and WT/DS48/AB/R, EC Measures Concerning Meat and Meat Products (Hormones). Report of the Appellate Body, AB-1997-4. Geneva, Switzerland: WTO.

9

Case Study 2: Plant Quarantines and Hass Avocados

ROLE OF SCIENCE IN SOLVING PEST QUARANTINE PROBLEMS: HASS AVOCADO CASE STUDY

WALTHER ENKERLIN HOEFLICH
Dirección General de Sanidad Vegetal, Secretaría de Agricultura,
Ganadería y Desarrollo Rural, México

Free trade among nations and regions is the driving force behind the relatively new and more widely accepted pest quarantine concept. In general, the quarantine concept for economically important pests has evolved from a near-zero threshold or near-zero pest tolerance, including the very stringent quarantine concept known as "absolute quarantine" to a more flexible approach in which a threshold above zero is allowed. This approach focuses more on integrating, in a system, a number of control measures in orchards, packing facilities, and transport to prevent pest establishment in pest-free countries or regions. This concept is known as a systems approach.

This modern approach facilitates trade by being less restrictive and at the same time providing quarantine security for the importing country. It is a more scientific approach that requires a comprehensive understanding of the pest biology and ecology as well as a larger and more solid infrastructure for the systems approach implementation. The Hass avocado case provides a good example on how science can be used to solve an old quarantine problem between two countries.

The state of Michoacan in the South Pacific coast of Mexico is a large producer of avocados. Michoacan grows around 100,000 ha of Hass avocados producing around 800,000 tons of fruit per year (Paz Vega, 1989). About 93 percent of the production is sold domestically and 7 percent is sold abroad. For the past 10 years the main importers of Mexican Hass avocados have been Japan, France, England, Switzerland, and Canada. The United States has been importing increasing amounts since 1997.

For over 80 years Mexico had been trying to export Hass avocados to the United States. However, exports were prohibited due to a quarantine restriction against three fruit fly species and two avocado fruit borers. The approval and enforcement of the North American Free Trade Agreement in 1991 provided space for negotiations and an opportunity for science to take part in the decision-making process (Figure 9-1).

The following section describes the experimental procedures used to solve this quarantine problem.

METHODOLOGY

For over 80 years the U.S. Department of Agriculture (USDA) had imposed a quarantine restriction on the Mexican Hass avocados, which are considered to be a host of the Mexican fruit fly (*Anastrepha ludens*, Loew), the sapote fruit fly (*A. serpentina*, Wied.), and the guava fruit fly (*A. striata*, Schiner), as well as a host of two species of avocado fruit borers.

Although it is known that some species of fruit flies infest certain avocado varieties (e.g., Sharwil avocado is considered to be a poor host of the Oriental fruit fly [*Bractocera dorsalis*] in Hawaii [Oi and Mau, 1989]), there is no scientific evidence of Hass avocado infestations by any fruit fly of the genus *Anastrepha*. In the case of the fruit borers it has been well documented that the Hass avocado is a primary host of this insect pest. However, scientific evidence also shows that fruit borers are temperature sensitive and their geographical distribution is restricted to certain altitudes within the avocado growing region in Michoacan. The avocado producing region in Michoacan is located at an altitude of 1400–2100 m above sea level and it is considered to have a temperate climate (Paz Vega, 1989). During the fall and winter months (October to February), the minimum and maximum temperatures fluctuate from 0 to 10 °C and from 16 to 20 °C, respectively.

A research project was conducted in Michoacan to assess the susceptibility of Hass avocados to the three above-mentioned fruit fly species and to determine the geographical distribution of the avocado fruit borers.

Considering that the quarantine problem implicated two countries, the exporter (Mexico) and the importer (United States), the research was approached in a binational fashion. The Mexican and U.S. governments decided to integrate a binational research team with fruit fly and quarantine specialists from both countries. The Mexican group Secretaria de Agricultura, Ganaderia y Desarrollo Rural, Direccion General de Sanidad Vegetal (SAGAR/DGSV) prepared a preliminary research protocol that was then sent to the USDA group of Animal and Plant Health Inspection Service/Agricultural Research Service (APHIS/ARS) for review. Once the research protocol was ready, a 10-month laboratory and field research program began in Michoacan. Periodical site visits were conducted by the USDA group for review and advice on the experiment. The research project was split into two main experiments and two side experiments as follows (see Figure 9-1). The main experiments were to determine the susceptibility of Hass avocados to *Anastrepha* spp. under forced laboratory conditions, and the susceptibility of Hass avocado to *Anastrepha* spp. under forced field conditions. The side experiments were to assess the *Anastrepha* spp. adult population fluctuation in the Hass avocado growing region, and the Hass avocado *Anastrepha* spp. natural field infestations.

For the laboratory and field susceptibility experiments, a range of physiological stages of avocado fruits were exposed to *Anastrepha* spp. forced infestations. The parameter used to measure the physiological stage of the fruits was percent dry matter (Enkerlin et al., 1994). The percent dry matter values used were 15, 17, 20, 22, 24, 26, 28, 30, 32, 34, and 35. Fruits containing less than 21 percent dry matter are considered to be unripe and with 21 percent or more are considered physiologically mature and ready for harvest.

For each percent dry matter, 40 avocado fruits were exposed immediately after harvest to forced infestations. Also, for each percent dry matter, fruits were exposed to forced infestations in the field while still attached to the tree and to laboratory forced infestations 3, 24, 48, 72, 96, and 120 hours after harvest. For the laboratory forced infestations, 50 sexually mature fruit fly couples were placed in 50-cm^3 cages. For the field experiments the same amount of couples were placed in 1-m^3 cages. In the laboratory, environmental conditions (temperature and relative humidity) were controlled to avoid detrimental effects on the fruit fly colonies used for the infestations. For the laboratory experiments, 10 fruits were placed per cage and were exposed for 24–48 hours to males and gravid females. For the field experiments fruits were exposed for 96 hours. After the infestation period, fruits were taken out of the cages and placed in plastic trays containing a 3-cm layer of vermiculite. Fruit was held for 18–25 days to allow for larvae development and pupation. After this time period fruits were dissected and vermiculite sieved to collect third instar larvae and pupae. The pupal stage was used as the critical developmental stage to assess Hass avocado fruit fly host status. The number of larvae and pupae was quantified and percent pupation and adult eclosion were calculated. The number of pupae per fruit was estimated to assess the severity of the infestations (Enkerlin et al., 1994).

Orange (*Citrus sinensis*), sapote (*Calocarpum sapota*), and guava (*Psidium guava*), which are the primary hosts of the fruit flies utilized in the experiment, were used as controls. For each of the avocado percent dry matter evaluated, 120 fruits of each primary host were exposed to fruit fly infestations. To compare the level of infestations of the fruit flies in their natural hosts against the avocado infestations and to be able to monitor the quality of the fruit fly colonies used in the experiment, the level of infestation (pupae/fruit) and

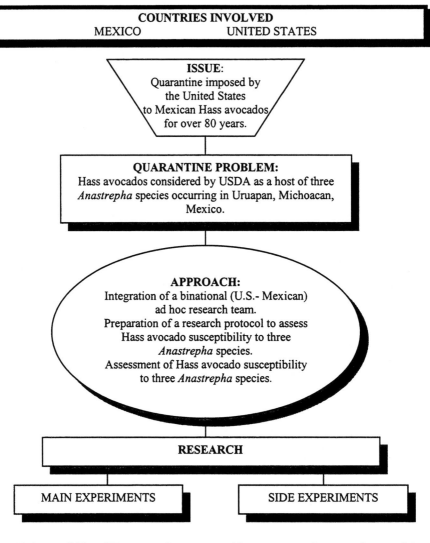

FIGURE 9-1. Case Study: Hass Avocado Background Chart. Binational research to assess the susceptibility of Hass avocados to pest infestation.

percent pupation of the three fruit fly species in their natural hosts was measured (Enkerlin et al., 1994).

The levels of infestation (pupae/fruit) obtained for each percent dry matter were statistically analyzed using an analysis of variance ($P = 0.05$) and a Tuckey studentized range test (SAS Institute, 1988).

To determine the presence and abundance of the three fruit fly species under study, an extensive network of McPhail traps was deployed and operated from July 1993 to April 1994 in the avocado growing region of Michoacan. A trap density ranging from one trap for every 1–5 ha was used. These trap levels meet with the protocol requirements recommended for certification programs in Mexico and the United States. Traps were serviced weekly and fly captures recorded. To obtain a measure of the fruit fly population levels, data were transformed to the population index, fly per trap per day (FTD). Also during the same time period, to assess natural fruit fly infestations, a systematic fruit sampling was conducted in the orchards where traps had been placed (Enkerlin et al., 1994).

RESULTS AND DISCUSSION

Forced Infestations

Under laboratory and field forced infestations, Hass avocados were shown to be a good host of *A. ludens*. Healthy *A. ludens* pupae were recovered from all the Hass avocado physiological stages that were evaluated. Hass avocados were a poor host at 15 and 17 percent dry matter. However, once the fruits approached physiological maturity (21 percent dry matter), they became highly susceptible (Figure 9-2, Table 9-1). For *A serpentina,* the Hass avocado becomes susceptible at 20 percent dry matter. Infestation levels for this species were lower but can still be considered a susceptible host especially at 20 and 22 percent dry matter. For *A striata,* a very low infestation was obtained at 22, 24, and 26 percent dry matter. Hass avocados are considered to be a very poor host of this fruit fly species (Figure 9-2, Table 9-1). The infestation drop observed at 24 percent dry matter is related to the fall and winter temperatures and photoperiod, not to an effect of the dry matter content on the development of immature stages of the insect. Once the temperature starts to rise and days become longer, infestation levels start to increase (Figure 9-2, Table 9-1) (Enkerlin et al., 1994).

In relation to the infestation on fruits that were attached to the tree and on those infested at different time intervals after harvest, results show, in general, an increase in susceptibility for each increase in time after harvest (Table 9-2). It is important to note that while the fruit was attached to the tree, at any physiological stage (percent of dry matter), no infestation was recorded. Twenty fruits were dissected to determine if female fruit flies had actually laid eggs under the skin of the fruit. Forty-eight egg masses (ca. 839 eggs) were found submerged in the fruit flesh with no signs of eclosion. Hard tissue surrounding

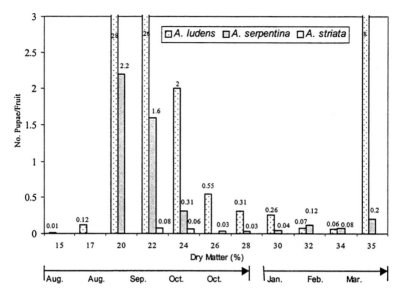

FIGURE 9-2. *A. ludens*, *A. serpentina*, and *A. striata* Forced Laboratory Infestation of Hass Avocado (Uruapan, Michoacan, 1993–1994)

TABLE 9-1. Statistical Analysis of Hass Avocado Fruit Fly Infestation Levels in Relation to Percentage of Dry Matter (Uruapan, Michoacan, 1993–1994)

Dry Matter[a] (%)	*A. ludens* Pupae/Fruit (avg.)	Statistical Difference[b] ($P = 0.05$)	*A. serpentina* Pupae/Fruit (avg.)	Statistical Difference[b] ($P = 0.05$)
15.20	0.02	B	0.00	B
17.20	0.10	B	0.00	B
20.20	26.00	A	1.70	A
21.50	24.00	A	1.20	AB
23.90	1.60	B	0.30	AB
26.40	0.50	B	0.00	B
29.50	0.30	B	0.03	B
30.80	0.20	B	0.05	B
32.30	0.64	B	0.10	B
33.40	0.57	B	0.08	B
35.00	7.46	B	0.15	B

[a]Parameter used in the experiment as an indicator of physiological maturity of avocado fruits: the higher percent dry matter, the more mature the fruit. (Alternatively, avocado oil content could be used as indicator of maturity.)

[b]The analysis of variance shows if there is statistical differences in infestation levels among percent dry matter values. Figures with the same letters are statistically equal. $P = 0.05$, Tuckey studentized range test (SAS Institute, 1998).

the egg masses in the form of callus was found (Enkerlin et al., 1994). This was also observed by Smith (1973), Sarooshi et al. (1979), and Armstrong et al. (1983) in previous experiments with other avocado varieties. These authors report this event to be a resistance factor of the avocado fruit associated to other unknown biological factors (Table 9-2).

Natural Infestations

Results from the 10-month trapping in the avocado region showed that, during this trapping period, only the Mexican fruit fly (*A. ludens*) was present in the region at detectable levels. The other two species (*A. serpentina* and *A. striata*) were not detected during the 10 months of trapping. *A. ludens* was captured throughout the trapping period except for January when the lowest temperatures were recorded in the region. *A. ludens* populations were detected at very low levels from July to February. During this time period population levels were below the low incidence threshold used by the Mexican Fruit Fly Campaign (CNCMF) as an action threshold indicating that populations are low enough to start a sterile fly release program for eradication purposes (CNCMF, 1993). Populations peaked during the months of March and April when temperatures in the region started to increase (Figures 9-3 and 9-4). Despite the population peak, levels are still low enough and do not exercise enough pressure to infest avocado fruits which are not the natural hosts of this or any other *Anastrepha* species (Hernandez-Ortiz, 1992). Furthermore, as Figure 9-3 shows Hass avocados are harvested throughout the year. However, the main harvest season is from October to May when around 60 percent of the total production is harvested (Paz Vega, 1989). During most of this time (October to February), *A. ludens* population levels are below 0.01 FTD, day which is the low-incidence threshold (Enkerlin et al., 1994).

Throughout the experiment, avocado fruits were sampled from the orchards where MacPhail traps had been placed. Fruits were systematically gathered from the ground and from the tree. A total of 2,311 kg (12,638 fruits) were gathered and dissected. No eggs or larvae were ever found in the fruits. Moreover, during the harvest season Plant Protection Official Inspectors assigned to agricultural districts 087 and 088 in Michoacan (which cover practically all the avocado growing regions sampled) sampled, in packing facilities, around 101 tons (405,534 fruits) with negative results in detection of immature stages of the pest (Santiago et al., 1994).

As the results clearly show, science provided basic information to assess, through a pest risk analysis, the feasibility of exporting Hass avocado fruits to the United States without jeopardizing the fruit industry in that country. Furthermore, it also provided the information required to mitigate risk, allowing for a systems approach implementation. Figure 9-5 schematically illustrates the role of science in solving a quarantine problem. It also shows how the information produced by the experiment, referred to here as a biological event,

TABLE 9-2. Susceptibility of Hass Avocados to *A. ludens*, *A. serpentina*, and *A. striata*—Forced Infestation in Fruits Attached to the Tree and in Fruits at Different Time Intervals after Harvest (Uruapan, Michoacan, 1993–1994)

Fruit Infestation after Harvest (h)	A. ludens			A. serpentina			A. striata		
	No. Fruits	No. Pupae	Pupae/ Fruit	No. Fruits	No. Pupae	Pupae/ Fruit	No. Fruits	No. Pupae	Pupae/ Fruit
0[a]	320	0	0	320	0	0	320	0	0
3	220	73	0.3	220	12	0.05	220	0	0
24	220	383	1.7	220	6	0.06	220	0	0
48	220	1,094	4.9	220	13	0.06	220	6	0.3
72	220	907	4.1	220	66	0.3	220	6	0.3
96	220	2,190	9.9	220	155	0.7	220	4	0.01
120	220	1,732	7.9	220	154	0.7	220	5	0.02

[a]Fruits subjected to infestation on the tree.

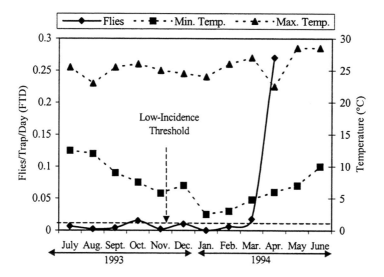

FIGURE 9-3. Seasonal Fluctuation of *Anastrepha ludens* Populations, and Minimum and Maximum Temperatures in the Hass Avocado Production Region of Michoacan, 1993–1994

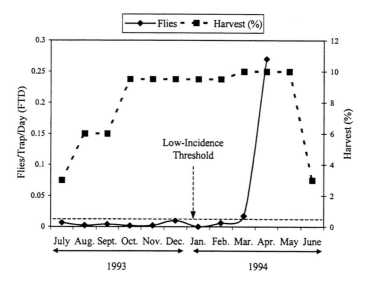

FIGURE 9-4. Seasonal Fluctuation of *Anastrepha ludens* Populations and Hass Avocado Harvest Period in Michoacan, 1993–1994

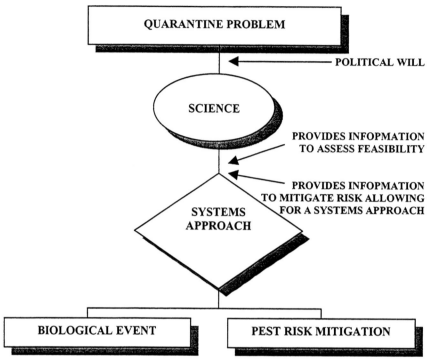

1. Natural resistance while attached to the tree.

1.1. None.

2. Susceptibility of Hass avocado fruits to *A. ludens* and *A. serpentina* forced infestations after harvest.

2.1. Sealed transport while in transit through Mexico.

2.2. Avocados not packed by the end of the work day must be kept in the screened fruit fly-proof packing area.

3. Presence of Mexican fruit fly in the avocado growing region of Michoacan.

3.1. McPhail trapping at protocol levels.

3.2. Control actions to keep populations at low-incidence levels.

4. Mexican fruit fly picks during the months of March–June.

4.1. Export season limited to November, December, and January.

FIGURE 9-5. Role of Science in Solving Quarantine Pest Problems

was used in a systems approach program to mitigate the risk of pest establishment associated with moving Hass avocados to the United States.

CONCLUSIONS

Under laboratory and field forced infestations, Hass avocado fruits are a good host *of A. ludens*, an average host of *A. serpentina*, and a poor host of *A. striata*. Hass avocado fruits attached to the tree are resistant to forced infestations of the three *Anastrepha* spp. evaluated. Hass avocado fruits are also resistant to infestations under natural field conditions The resistance observed is biochemical and ecological, not physical or morphological.

GENERAL CONSIDERATIONS

• Biological sciences should always play a key role in solving pest quarantine problems. Science provides information to assess feasibility based on levels of risk, and it also provides information to mitigate risk.
• The scientific approach should include a multidisciplinary group of scientists in order to have a comprehensive understanding of pest-host relations.
• Evaluations of natural field infestations should be mandatory over laboratory and field forced infestations for pest risk assessment.
• Putting together a binational research team and a follow-up through site visits were key components on reaching the experimental goals.
• The overall scheme used in solving the Hass avocado quarantine dispute can be considered an effective model in solving quarantine problems.

REFERENCES

Armstrong, J.W., W.C. Mitchell and G.J. Farias. 1983. Resistance of "Sharwil" avocados at harvest maturity to infestations by three fruit fly species (Diptera:Tephritidae) in Hawaii. Journal of Economic Entomology 76:119–121.
CNCMF (Campaña Nacional Contra Moscas de la Fruta). 1993. Manual de opraciones de campo. Documento oficial no publicado. Dirección General de Sanidad Vegetal SAGAR. Coyoacán, México.
Enkerlin H.W., J. Reyes F., A. Bernabe A., J.L. Sanchez P., J. Toledo A, and M. Aluja S. 1994. Estatus del aguacate Hass como hospedero de tres especies de moscas de la fruta del género Anastrepha, (Diptera:Tephritidae), bajo condiciones forzadas en laboratorio y campo, y bajo condiciones naturales en campo. Campaña Nacional Contra Moscas de la Fruta. Dirección General de Sanidad Vegetal, SARH. Agrociencia serie Protección Vegetal 4(3):230–348.
Hernández-Ortiz, V. 1992. El género Anastrepha Schiner en México (Diptera:Tephritidae)., Veracruz, México: Instituto de Ecología, A.C., Xalapa.
Oi, D.H., and R.F. Mau. 1989. Relationship of fruit ripeness to infestation in "Sharwil" avocados by Mediterranean fruit fly and Oriental fruit fly (Diptera:Tephritidae). Journal of Economic Entomology 82:556–560.

Paz Vega, R. 1989. Mexican avocados: threat or opportunity for California? California Avocado Society, pp. 87–106.
Santiago M., G., W. Enkerlin H., J. Reyes F.,and V. Ortiz G. 1993. Ausencia de infestación natural de moscas de la fruta (Diptera:Tephritidae) en aguacate Hass en Michoacán, México. Campaña Nacional Contra Moscas de la Fruta. Dirección General de Sanidad Vegetal, SARH. Agrociencia serie Protección Vegetal 4(3):349–357.
Sarooshi, R.A., D.R. Blundell, and D.L. Peasly. 1979. Blemish and abnormalities of avocado fruit. Agricultural Gazette of New South Wales 90:18–20.
SAS Institute. 1988. SAS User's Guide, release 6.03 ed. Cary, North Carolina: SAS Institute.
Smith, D. 1973. Insect pests of avocados. Queensland Agricultural Journal 99:645–653.

THE HASS AVOCADO CASE:
A POLITICAL SCIENCE PERSPECTIVE

DAVID VOGEL

Haas School of Business, University of California, Berkeley

Since 1914, the import of the Hass avocado from Mexico into the United States had been forbidden on the grounds that the fruit was a host of various fruit flies and seed pests whose introduction into the United States would threaten the American avocado crop. Recently, this ban has been lifted. Although still subject to various restrictions, Mexican avocados can now be exported. At one level, this significant policy shift reflects the development and implementation of a set of scientific protocols and procedures that have provided assurance that the fruit sold in the United States does not contain these harmful pests. But although scientific arguments and evidence may have been a necessary condition for trade liberalization, they were certainly not a sufficient condition. It was politics, in the form of the North American Free Trade Agreement (NAFTA) that provided the necessary impetus for the ending of an embargo after more than 80 years (Vogel, 1995).

For at least two decades Mexico had been pressuring the United States to come up with a protocol that would allow the importation of Mexican avocados. A number of specific proposals were explored during the 1970s, and considerable progress appeared to have been made in developing procedures that would provide adequate protection to American growers. At one point, the U.S. Department of Agriculture (USDA) was considering proposing a rule in the *Federal Register* that would have made it possible to begin exploring the implementation of these procedures. However, the California Avocado

Commission, with the assistance of California Senator Cranston, persuaded the USDA not to issue such a rule.

Negotiations between the United States and Mexico continued and there was a good deal of additional scientific research, but the considerable political influence of the California avocado industry effectively prevented the adoption of any protocols or procedures that would have relaxed the embargo and thus exposed California growers to international competition. This represented a major setback for Mexican growers who had made substantial progress in devising various means of preventing exports of their crop from endangering American agriculture.

During the intense domestic debate over the adoption of a free trade agreement with Mexico, major segments of American agriculture opposed congressional approval of NAFTA for straightforward protectionist reasons: They did not want to have to compete with less expensive Mexican agricultural exports. Among the agricultural producers opposed to NAFTA was the California avocado industry who clearly understood that the approval of this trade agreement would reduce their ability to keep out lower-priced Mexican produce and thus reduce their profits or market share.

NAFTA was, of course, approved. Not surprisingly, among the first requests from Mexico following its approval was for an ending of the American restriction on exports of avocados. After some delays, the USDA issued new regulations that ended the embargo, and Mexican avocados are now available for sale in the United States under various conditions.

In effect, what NAFTA did was to end the monopoly of California avocado growers over policy making at the USDA. Prior to NAFTA, there was no domestic constituency that favored the relaxation of import controls. To be sure, such a relaxation was clearly in the economic interests of American avocado consumers, but they were not politically organized. No American consumer group chose to focus on this issue and it received little or no press coverage. The typical American consumer neither knew nor cared that they were paying above world market prices for avocados, and as a result their interests were not represented in the policy process.

What NAFTA did was to give political voice to a constituency that favored the relaxation of import controls, namely Mexican avocado growers. Trade agreements, by definition, globalize domestic politics: They explicitly give foreign producers a claim on domestic policy making. At the same time, trade liberalization also serves to give foreign producers domestic allies. With the approval of NAFTA, American producers now have the opportunity to gain access to the Mexican market. These producers now have the ability to demand that Mexico eliminate or reduce its use of sanitary and phytosanitary (SPS) standards to keep out their products. Thus the avocado agreement can be seen as part of a broader, reciprocal agreement to reduce the use of SPS standards as trade barriers on both sides of the border—a dynamic that NAFTA made possible.

The approval of NAFTA served to politically isolate California avocado producers. While NAFTA was being negotiated, their interests coincided with those of other American growers who opposed trade liberalization with Mexico. But once NAFTA was approved, each crop was on its own. Significantly, in their last-ditch effort to prevent the opening of the American markets, the California avocado growers received only token backing from other growers, each of whom was now focused on protecting their own markets.

The liberalization of the American avocado market represents an almost textbook case of the benefits of trade liberalization. For what NAFTA did was to subject American "scientific" restrictions on avocado imports to international scrutiny. And it turns out that they were unable to survive such scrutiny. In effect, NAFTA made possible the triumph of science over economics. Without NAFTA, or more specifically, the access that NAFTA accorded the scientific claims of those who favored trade liberalization, the various scientific protocols and procedures that had been devised to permit the importation of Mexican avocados in ways that did not endanger American growers would have remained stillborn. From this perspective, this case illustrates trade liberalization at its best: It changed a regulation whose only purpose was to protect the economic interests of American producers, and, as a result of this change, American consumers are better off.

At the same time, it is important to keep the significance of this case study in perspective. This dispute over SPS standards was primarily an economic one: It pitted the interests of American avocado growers against Mexican growers. There was no question of consumer safety; what was in dispute was the "safety" of domestic growers. American producers and consumers had opposite interests. Because the former could not claim that the latter's health and safety would be endangered if the importation of Mexican avocados was permitted, American consumer groups could not be mobilized to back the avocado ban. This stands in sharp contrast to, for example, the beef hormone dispute between the United States and the European Union, which does raise politically salient consumer health issues. Accordingly, European consumers and consumer groups have become important allies of protectionist producers. This makes the resolution of this dispute through appeals to "science" much more difficult.

REFERENCE

Vogel, D. 1995. Trading Up: Consumer and Environmental Regulations in a Global Economy. Cambridge, Mass.: Harvard University Press.

10

Case Study 3: Genetically Modified Organisms

AN OVERVIEW OF RISK ASSESSMENT PROCEDURES APPLIED TO GENETICALLY ENGINEERED CROPS

PETER KAREIVA and MICHELLE MARVIER
Department of Zoology, University of Washington

The commercial production of genetically engineered crops has prompted countries around the world to adopt risk assessment procedures for evaluating the safety of transgenic cultivars. Most concern has been directed at the risk that a genetically modified crop may itself be made more weedy as a result of its recombinant trait, or may, through hybridization and introgression, contribute genes to a wild relative, consequently making the related plant more weedy (reviewed in Williamson, 1993; Rissler and Mellon, 1996; Bergelson et al., in press). Additional risks include the environmental fate of plant products (such as degradation versus accumulation of novel endotoxins in soils) and altered agricultural practices (such as increased application of herbicides; Rissler and Mellon, 1996). Although these ecological risks are widely thought to be on average minimal, the tremendous variety of plant attributes that are potentially modifiable renders blanket pronouncements of safety untenable. Moreover, because experience with transgenic crops is still limited, the formal development of risk assessment procedures faces the challenge of anticipating problems with traits that have not yet been developed let alone patented or commercialized.

In spite of striking cultural differences regarding willingness to accept risk, countries around the world have converged on three general principles of risk assessment for transgenic crops: containment, the principle of familiarity, and a reliance on small-scale experiments. We discuss each of these approaches and their limitations. Finally, in recognition of the shortcomings of existing screening procedures, we end with a recommendation that greater consideration be given to postrelease monitoring of transgenic plantings.

CONTAINMENT

The most straightforward way to manage the risk of a biological organism would be to simply contain the organism, to somehow prevent it from spreading beyond its intended release site. For instance, the initial experiments with genetically engineered ice-minus bacteria in Northern California were subjected to elaborate security measures, including fences and broad isolation zones. In its 1989 report on the field testing of genetically modified organisms, the National Research Council (NRC) offered the optimistic conclusion that "routinely used methods for plant confinement offer a variety of options for limiting both gene transfer by pollen and direct escape of the genetically modified plant" (NRC, 1989, p. 36). If transgenic plants and genes could in fact be contained, decisions regarding their risks would be greatly simplified. Yet, on the contrary, data from field trials clearly demonstrate that this initial faith in the feasibility of containment was overly optimistic. For some species hybridization and transfer of genes to wild relatives can occur very rapidly (e.g., Mikkelsen et al., 1996). In addition, direct field experiments indicate that, although most pollen moves only short distances from source plants, a measurable quantity of pollen travels vast distances, making containment of transgenic pollen highly unlikely (e.g., Kareiva et al., 1991; Kareiva et al., 1994; Lavigne et al., 1998).

Potential methods of containment include the use of barren zones around crops and plantings of trap plants into border rows. Unfortunately, barren zones may actually cause increases in the mean distance or amount of gene flow out of plots (Manasse, 1992; Morris et al., 1994). Although the use of border rows to trap pollen has proven more successful in reducing the extent of gene movement, the borders must be substantially larger than the transgenic fields, making their use impractical for agronomic-scale plantings (Hokanson et al., 1997).

Even in cases where gene transfer is an extremely infrequent event, the notion that transgenes could ever be completely contained remains indefensible. Furthermore, with large-scale commercial production, the sources of transgenes are so plentiful and opportunities for exchange so widespread, containment can not possibly be considered as a tenable risk management procedure. It is noteworthy that regulations in the United States and in the European Union do not in any way rely on containment as part of their risk management procedure for commercial products. In these countries, containment practices are only required for small-scale experiments during the research and development stage of novel cultivar breeding and genetic modification.

THE PRINCIPLE OF FAMILIARITY

Risk assessments often rely on comparisons between transgenic plants and the more familiar unmodified form of the plant or closely related plant species. The Organization for Economic Cooperation and Development (OECD) describes this principle as follows:

> Whether standard cultural practices would be adequate to manage a relatively unfamiliar new plant line or cultivar can be assessed based on familiarity with a closely related line in conjunction with results from laboratory and preliminary field work with the new line (Anonymous, 1993).

This principle is not intended to imply that "familiarity means safety," although implementation of the policy frequently seems to embody such a deduction. For example, it is often assumed that if experiences with familiar plants have been broad and generally positive (e.g., the unmodified plant and its close relatives are not weeds), then the transgenic plant is similarly unlikely to pose a substantial risk. However, field experiments have clearly demonstrated that genetic modification may result in a number of incidental changes to the plant's original traits and that extrapolations from the familiar to the unfamiliar can be severely misguided. For example, the common weed *Arabadopsis thaliana* is a highly selfing species for which the prospects of gene transfer would generally be considered very low. However, field experiments with transgenic *Arabadopsis* showed that the transgenic plants, for some unknown reason, actually outcrossed at a rate of 6 percent, nearly 20 times more frequently than unmodified *Arabadopsis* (Bergelson et al., 1998). The authors concluded (p. 25) that "genetic engineering can substantially increase the probability of transgene escape, even in a species considered to be almost completely selfing." Although regulations in some nations advise that the required degree of scrutiny should depend on the traits of the parent organism (e.g., Genetic Manipulation Advisory Committee, 1998, Appendix 5), transgenic plants may exhibit substantially altered life histories and "familiarity with these [parental] species as useful agricultural and horticultural plants may be irrelevant and misleading" (Williamson, 1994).

A second problem regarding application of the principle of familiarity arises when the risk of a recombinant trait is compared with that of a familiar, seemingly similar trait that occurs naturally in unmodified plants. The assumption is that a novel trait that is similar to traits seen elsewhere is unlikely to pose new risks. The problem is that familiarity with a trait is in the eyes of the beholder. An especially good example involves the gene derived from *Bacillus thurengiensis* (Bt) for endotoxin production, which provides a "natural" insecticide. Because plants in general produce compounds that act as antiherbivore agents, and plant breeders have a long tradition of selecting plant varieties to increase their resistance to herbivores, some might argue that Bt endotoxin production is "familiar" and therefore probably "safe." On the other hand, when the gene for Bt endotoxin is inserted into canola, the transgenic canola acquires a trait that it has never before possessed; a trait that protects it, to varying degrees, from a very broad range of caterpillar species. The risks associated with such a trait should not be assessed on the basis of subjective opinions regarding its familiarity or novelty, but rather should rely on data from experimental trials.

A third type of extrapolation that is tenuous concerns the long-term effects of repeated plantings of genetically modified crops on soil ecosystems. For example, although Bt endotoxins have previously been sprayed on crops as a form of organic pest control, we have no experience with large quantities of Bt-laden crops decomposing in soils year after year. Experiments have indicated that Bt-residues in cotton leaves persisted for at least 56 days after burial in the soil (Palm et al., 1996). Similarly, although small-scale laboratory experiments indicate no harmful impacts of proteinase inhibitors (another transgenic trait with insecticidal activity), longer-term experiments using natural soil communities suggests that there might be surprising impacts of these compounds with respect to microbial respiration and soil organisms (Donegan et al., 1997).

Extrapolations from the familiar to the unfamiliar of the type described above are common, but improper, applications of the principle of familiarity. Rather, the intention of the principle is that familiarity should provide a context for measuring risk—for example, the weediness of a genetically modified plant could be compared with that of the familiar, unmodified form. In fact, U.S. regulations require that before a transgenic crop is deregulated, it must be shown that the genetically engineered plant "is unlikely to pose a greater plant pest risk than the unmodified organisms from which it was derived" (U.S. Department of Agriculture [USDA], 1992). Although surprisingly few of the U.S. petitions for nonregulated status approved prior to 1995 performed such a comparison (Purrington and Bergelson, 1995), experiments comparing the performance of transgenic plants with unmodified source plants should be a cornerstone of the risk assessment process. Thus, rather than providing any evidence regarding risk, familiar plants should provide a benchmark or standard to which the risks posed by modified plants can be compared.

SMALL-SCALE RISK ASSESSMENT EXPERIMENTS

Most countries require some degree of "testing" to quantify risks if a crop is modified in a way that seems ecologically significant. In the United States, the earliest petitions to deregulate transgenic crops tended to be deficient on actual field experiments and instead relied upon greenhouse tests or simple literature surveys (Parker and Kareiva, 1996, Table 1). Although disputes have arisen repeatedly between environmental groups and industry over the appropriateness of various experimental designs (e.g., Rissler and Mellon, 1996, comment on Upjohn's transgenic squash petition, Animal and Plant Inspection Service [APHIS] Docket No. 92-127-1) and experimental risk assessments have generally been severely flawed (Purrington and Bergelson, 1995), reliance upon field experiments has grown steadily over recent years. Currently, in the United

States, Europe, and Australia, field experiments aimed at evaluating the potential weediness of transgenic crops are a mandatory part of the approval process (USDA, 1992; European Communities Committee, 1998; Genetic Manipulation Advisory Committee, 1998).

Field experiments are, in fact, a valuable tool: If a transgenic crop behaved like an aggressive weed in these experiments, it would be a clear signal that the plant should be tightly regulated and perhaps not allowed for commercial production. However, while the experimental detection of weediness provides a clear sign of danger, the failure to detect weediness does not lead to such a clear-cut conclusion. Determination of "safety" is more complicated because we must consider the experiment's capacity to detect weediness if it in fact exists. Unfortunately, a one- to two-year field assessment in small plots over a limited region may fail to reveal any enhancement of weediness, when in fact such an enhancement occurs under infrequent but important conditions. Simulations demonstrate that field tests for assessing a plant's enhanced invasiveness are prone to high rates of error unless the trials are repeated at multiple sites and over at least several years (Kareiva et al., 1996). Similarly, the potential risks associated with herbivore resistance genes can only be assessed accurately when trials are performed at multiple sites that offer potentially different environments for plant growth as well as different background densities of herbivores (Marvier and Kareiva, 1999).

A further weakness of short-term experiments is that there will likely be substantial time lags between the introduction of a transgenic plant and the emergence of ecological problems related to its introduction, such as escape of transgenes into wild relatives or the naturalization of transgenic crops. Long time lags are inherent features of many biological invasions. For example, a survey of historical records for past invasions by weeds in the northwestern United States indicated that the median timelag between the first record of a weed and the onset of widespread infestation was on the order of 30–50 years (Marvier et al., 1999). In addition, time lags between the introduction of ornamental woody plants and their escape into the wild in Germany are on the order of 150 years (Kowarik, 1995). Although examples from the "exotic species" literature are often rejected in the biotechnology arena, it is entirely reasonable to expect that invasions of transgenes will entail extensive time lags simply because invasion is such an unlikely event, probably depending on the chance concordance of a suite of favorable conditions. The potential for time lags means that short-term experiments are likely to support a verdict of "safety" when in fact such a determination is not warranted.

MONITORING AND A PRECAUTIONARY APPROACH

Unfortunately, containment of transgenic plants or their genes is not a viable option, "familiarity" with related plants or similar traits cannot be extrapolated accurately to the transgenic plants themselves, and a few experiments under a narrow range of conditions can not provide acceptable proof of safety. In light of the tremendous uncertainty of risk assessment, the European community has called for amendments to Directive 90/220/EEC on deliberate release of genetically modified organisms that would require vigilant monitoring of transgenic commercial plantings after a marketing consent has been granted (European Communities Committee, 1998), with the idea that dangerous escapes might be detected before undue damage has been done. This approach could prove feasible if populations of problematic transgenic crops (or transgenic weeds) might be sufficiently confined and then controlled with herbicide.

Long-term, large-scale monitoring of transgenic plantings provide both an important research opportunity—we can learn a great deal about temporal and spatial variability as well as the occurrence of rare events—and a valuable means of minimizing risk. Although caution and tenacious monitoring are clearly warranted for certain transgenic crops, it will be hard to exercise that caution given the current pressure to ease regulations on the basis of a safe record to date. It should, however, be considered that, although monitoring is an expensive enterprise, the cost and difficulty of controlling a weed population are greatly exacerbated once a weed becomes well established. Thus, investment in monitoring programs that strive toward the earliest possible detection and elimination of transgenic weeds will likely prove cost effective in the long run. More generally, a reliance on monitoring when uncertainty, in the face of empirical data, is still substantial may be an advisable principle for a wide variety of risk assessments. Because of evolution and the role of chance in biological dynamics, monitoring may need to be a mainstay of any ecological risk assessment.

REFERENCES

Anonymous. 1993. Safety consideration for biotechnology: Scale-up of crop plants. Paris: Organization for Economic Cooperation and Development (OECD).

Bergelson, J., C.B. Purrington, and G. Wichmann. 1998. Promiscuity in transgenic plants. Nature 395:25.

Bergelson, J., J. Winterer, and C.B. Purrington. In press. Ecological impacts of transgenic crops. In Biotechnology and Genetic Engineering of Plants. V. Malik, ed. Oxford, U.K.: Oxford University Press.

Donegan, K.K., R.J. Seidler, V.J. Fieland, D.L. Schaller, C.J. Palm, L.M. Ganio, D.M. Cardwell, and Y. Steinberger. 1997. Decomposition of genetically engineered tobacco under field conditions: persistence of the proteinase inhibitor I product and effects on soil microbial respiration and protozoa, nematode, and microarthropod populations. Journal of Applied Ecology 34:767–777.

European Communities Committee. 1998. Second Report: EC Regulation of Genetic Modification in Agriculture. 6378/98/98 (COM(98) 85) Proposal for a European Parliament and Council Directive amending Directive 90/220/EEC on the deliberate release into the environment of genetically modified organisms.

Genetic Manipulation Advisory Committee. 1998. Guidelines for the Deliberate Release of Genetically Manipultated Organisms: Field Trials and General Release. Canberra, Australia.

Hokanson, S. C., R. Grumet, and J. Hancock. 1997. Effect of border rows and trap/donor ratios on pollen-mediated gene movement. Ecological Applications 7:1075–1081.

Kareiva, P., R. Manasse, and W. Morris. 1991. Using models to integrate data from field trials and estimate risks of gene escape and gene spread. Pp. 31–42 in Biological Monitoring of Genetically Engineered Plants and Microbes. D. R. MacKenzie and S. C. Henry, eds. Bethesda, MD: Agricultural Research Institute.

Kareiva, P.W. Morris, and C.M. Jacobi. 1994. Studying and managing the risk of cross-fertilization between transgenic crops and wild relatives. Molecular ecology 3:15–21.

Kareiva, P., I.M. Parker, and M. Pascual. 1996. Can we use experiments and models in predicting the invasiveness of genetically engineered organisms? Ecology 77:1670–1675.

Kowarik, I. 1995. Time lags in biological invasions with regard to the success and failure of alien species. Pp. 15–38 in Plant Invasions: General Aspects and Special Problems. P. Pysek, K. Prach, M. Rejmanek, M. Wade, eds. Amsterdam: SPB Academic Publishing.

Lavigne, C., E.K. Klein, P. Vallee, J. Pierre, B. Godelle, and M. Renard. 1998. A pollen-dispersal experiment with transgenic oilseed rape. Estimation of the average pollen dispersal of an individual plant within a field. Theoretical and Applied Genetics 96:886–896.

Manasse, R. 1992. Ecological risks of transgenic plants: effects of spatial dispersion on gene flow. Ecological Applications 2:431–438.

Morris, W.F., P.M. Kareiva, and P.L. Raymer. 1994. Do barren zones and pollen traps reduce gene escape from transgenic crops? Ecological Applications 4:157–165.

Marvier, M.A. and P. Kareiva. 1999. Extrapolating from field experiments that remove herbivores to population-level effects of herbivore resistance transgenes. Pp. 57–64 in Proceedings of a Workshop on: Ecological Effects of Pest Resistance Genes in Managed Ecosystems. Traynor, P.L. and J.H. Westwood, eds. Blacksburg, Virginia: Information Systems for Biotechnology.

Marvier, M.A., E. Meir, and P.M. Kareiva. 1999. How do the design of monitoring and control strategies affect the chance of detecting and containing transgenic weeds? In Risks and Prospects of Transgenic Plants, Where Do We Go From Here? K. Ammann and Y. Jacot, eds. Basel: Birkhasuer Press.

Mikkelsen, T.R, B. Andersen, and R.B. Jorgensen. 1996. The risk of crop transgene spread. Nature 380:31.

National Research Council (NRC). 1989. Field Testing Genetically Modified Organisms: Framework for Decisions. Washington, DC: National Academy Press.

Palm, C.J., D.L. Schaller, K.K. Donegan, and R.J. Seidler. 1996. Persistence in soil of transgenic plant produced *Bacillus thurengiensis* var. kurstaki delta-endotoxin. Canadian Journal of Microbiology 42:1258–1262.

Parker, I.M. and P. Kareiva. 1996. Assessing the risks of invasion for genetically engineered plants: acceptable evidence and reasonable doubt. Biological Conservation 78:193–203.

Purrington, C.B. and J. Bergelson 1995. Assessing weediness of transgenic crops: industry plays plant ecologist. Trends in Ecology and Evolution 10:340–342.

Rissler, J. and M. Mellon. 1996. The Ecological Risks of Engineered Crops. Cambridge, MA: Massachusetts Institute of Technology Press.

U.S. Department of Agriculture (USDA). 1992. Federal Register 57, 53036–53043.

Williamson, M. 1993. Risks from the release of GMOs: ecological and evolutionary considerations. Environment Update 1:5–9.

Williamson, M. 1994. Community response to transgenic plant release: predictions from the British experience of invasive plants and feral crop plants. Molecular Ecology 3:75–79.

APPROACHES TO RISK AND RISK ASSESSMENT[1]

PAUL B. THOMPSON

Department of Philosophy, Purdue University

Risk analysis is typically understood as a wholly technical or scientific process. Yet the very concept of risk usually implies that some class of possible events has been judged to be adverse, or that that the very indeterminacy of future events is itself adverse. As such, risk analysis cannot be wholly based on science. At best, science can characterize the mechanisms that would lead to events such as mortality or morbidity, and can assign a probability or likelihood to their occurrence. Still, the badness or adversity that is associated with death and disease is based not on science, but morality. Nature is indifferent to death, and it is only when the perspective of human striving is introduced that it can be understood in terms of risk. Risks to health seem amenable to a purely scientific characterization because the moral judgments that are involved in this issue are among the least controversial. But even these judgments become contested at the margins. Ideas of "health" shift from "absence of disease" to "enhanced capacities," and the capacity to control (and hence assume responsibility for) future events is reflected in the judgment that a particular practice is "risky." As such, philosophy and ethical theory have an inevitable place in the characterization and evaluation of risks.

Within the social sciences, the normative and philosophical dimensions of risk are often incorporated into the characterization of rationality. For example, cost–benefit analysis (discussed in Chapter 2) frames rational choice through evaluating and comparing the likely outcomes from each of two or more options. Cost–benefit analysis takes on ethical significance when rational

---

[1]Author's note: The following is a lightly edited transcript of my workshop presentation, which was an overview of my own research as it bears on the case of genetically modified foods. It was not intended to be a comprehensive or representative discussion of philosophical work on risk assessment or on biotechnology. The orientation of the chapter is thus personal and citations are strongly biased toward my own publications. There has been an on-going discussion of this topic in popular press and on the Internet. Thompson (1997a) provides a more balanced and fully referenced discussion of philosophical work on biotechnology.

optimization of expected values is presumed to be the decision rule that should guide decision making with respect to regulatory standard setting or investment of public resources. Philosophical research on risk has tended to take one of two tacks with respect to this conception of rational optimization. Philosophers who endorse the basic strategy of rational optimization have tended to be critical of scientists' characterizations of probability and uncertainty (see Shrader-Frechette, 1991; Wachbroit, 1991). Other philosophers are critical of rational optimization and cost—benefit analysis, and have argued that public choices should focus on maintaining a basic structure of rights that maintain conditions of fairness among private decision makers (see Sagoff, 1985, MacLean, 1990).

For the case study presented by Peter Kareiva and Michelle Marvier, I will introduce a different set of philosophical concerns that focus on ways of framing (or interpreting) risks involved with genetically engineered food. One of the parameters that I use in my work is not to question the consensus assessment among scientists about the probability and degree of harm associated with genetic engineering. Sometimes it is difficult to figure out exactly what that consensus is, but to the extent that I can discern it, I never question it. That is not my business as a philosopher. What I am interested in is the divergence between that assessment, however it is set it up, and that of the broader public (or at least some segments of the broader public) with respect to the riskiness of genetically engineered food. There are, of course, differing opinions among scientists. Nonetheless, it has been and still is true that the broader public (and particularly if that is extended to the specifically concerned public) understand genetically engineered food to be riskier than the scientific consensus would suggest. My particular project has been to try to understand the rational basis for that difference. I am not interested in irrational bases for difference. I am not interested, for example, in purely nonrational judgments of taste. And in some sense, I am not even interested in culture as an explanatory value of those differences, although I do believe that culture has a tremendous influence in terms of the way that people understand risk and get information about risk.

I have been strongly influenced by cognitive work on risk undertaken by people such as Paul Slovic and, before that, Tversky and Kahnemann (1982). But unlike them, my framework is rational choice, and I am interested in the rational basis for deviations between a benchmark notion of what the risk is, derived from scientific consensus and other notions that might be held by the public. Furthermore, my project is a philosophical rather than an empirical one: I am attempting to make sense of the debate over genetically modified organisms in a manner that exposits and exemplifies a conception of rationality. I am not attempting to make empirical claims about human psychology or motivation. The philosophical work that I have done suggests testable empirical hypotheses, but I do not represent my work as making empirically verified claims.

My philosophical approach to the subject hand is non-standard in that I do not assume that probability and harm or probability and negative outcome are essential characteristics of risk. I have built my work on risk by looking at the way the word risk is actually used in Western languages (Thompson, 1987 and 1991; Thompson and Dean, 1996). I look for the meaning of the word "risk," the things that it could possibly mean in a grammatical sentence. Although in many instances it could and does, in fact, mean something like "the probability of harm," that certainly does not account for all of the legitimate uses of the word risk. So I would argue that we need a broader notion of risk, one that sees it as having multiple dimensions. This is a standard view in risk perception and cognitive science literature (Slovic, 1987).

My hypothesis is that although genetic engineering tends to score fairly low with respect to probability and harm, it tends to score fairly high with respect to some of these additional dimensions of risk. In this paper I discuss two dimensions of risk. One is information reliability and the second is an ambiguity between event-predicting and act-classifying notions of risk (see Thompson, 1997a; 1997b; 1999).

First, is information reliability. Whenever anyone does work on risk, one of the factors to be considered is how reliable the information is. We tend to discount information that we believe to be unreliable. In the first part of this chapter Peter Kareiva and Michelle Marvier discuss the value judgments that scientists apply within their research and within their community for how much discounting to place on information. Here I lay out a spectrum between highly reliable information that is true (although in some respects that is a bad, possibly misleading characterization), to highly unreliable information, which is not just false but also mendacious.

How do people sort out whether information is highly reliable or highly unreliable? Clearly one of the things that people consider in evaluating reliability is the context in which this information is presented to them. As a matter of fact, I would argue that the discourse context—the kind of speech that is being performed, the kind of claims that are being made, the purposes that are behind the making of claims, and the rules under which claims can be put forward and evaluated—all influence the extent to which people regard information as reliable. Corresponding to highly reliable information we can postulate the ideal discourse situation, which is a long story. It is something borrowed from the work of Habermas (1990). In the ideal discourse situation, everyone is trying to figure out what is true. There are rules of arguments and ethics; there are possibilities of reproducing results or testing results that are carried out. So there is a sense, at least, in which the way science is supposed to work that fits the ideal discourse situation, and it is clear that people like Habermas who have worked this out have science in mind when they talk about ideal discourse.

On the opposite extreme, there is strategic discourse, and purely strategic discourse is a situation in which people do not care about whether the claim is true or false. Strategic speakers only want you to believe something or to act on the basis of something or to accept it as true because it happens to suit some particular interest of theirs at the moment. My paradigm example of strategic discourse in some of my writings is buying a used car. Not all used car dealers are bad, of course, but the metaphor still strikes a chord. The used car dealer is a cultural icon—we just do not believe anything that a used car dealer tells us.

There is a rational tendency to regard a situation as more risky (like buying a used car) to the extent that we see it moving down a scale toward more strategic considerations and toward more circumstances in which the information that we get is expected to be unreliable. My conclusion would be that risk increases to the extent that one is moving down the information reliability scale. We tend to think of this as *the* risk of buying a car, which is risky. There is some sense in which the objective facts about the probability that the car is going to break down are quite independent of whether the person that is selling us the car is with a firm that we trust and so on. But we will interpret the purchase of the car and the activity of buying the car as more risky based in part on this information reliability factor.

So this is one dimension in which there is a tremendous difference between the public's position and the position of the scientific community, including the regulatory community. The difference is that, for the most part, the scientific community's information about risk comes from an ideal discourse situation. As scientists, we may not get quite as close to an ideal discourse as we might like in large conference settings, but it is far closer to an ideal discourse setting than the circumstance in which members of the public often acquire risk information. Therefore, it is, in fact, quite rational to regard information that filters through strategic channels as questionable. In other words, if genetic engineering is claimed to be safe in a strategic situation, someone might actually interpret that claim to mean that it is therefore more dangerous because it is claimed to be safe. If it is claimed to be dangerous in a strategic situation, one might actually move in the other direction and think that therefore it must be safe.

Again, I will not speculate too much on whether and how much this explains European versus North American considerations. But it may well be that there is a sense in which, partly because of the way in which the issue has come to Europe as part of the strategic trade negotiations, that there is a tendency to see this as a set of more strategic claims than in the United States.

The second issue that I want to point out is a bit more contentious and a bit more complex. There is an ambiguity in the concept of risk that I have characterized here, and I am systemizing it as an ambiguity between event predicting and act classifying. If we look at the way that people talk about risks in real life, in a nonscientific context, often what they mean is exactly what scientists mean, which is that some function of the probability of events, and the value or harm are associated with the events. But there are many other contexts in which that cannot be what is meant. To summarize a long argument (Thompson, 1991 and 1995), remember that the word "risk" is a verb. And words like "risky" and "risking" pertain much more to the verb form of the word risk than to the noun form of the word "risk". I defy anyone to translate probability and harm into a verb. When someone risks something, they are doing something. There is some connotation of action or activity that is implicit whenever the word risk is used as a verb. There is no connotation of action that is implicit when the word risk is used as a probability and an outcome.

Furthermore, if you'll perform the thought experiment, you will have a lot of trouble forming a meaningful grammatically correct English sentence in which the subject that risks, the subject of a risk sentence, is not an intentional agent. By that I mean a human being or a group. We attribute intentionality to corporations and countries all the time. Sometimes we attribute it to animals. We do not attribute it too often to plants and trees, and we certainly do not attribute it to mountains and ecosystems; it just does not make sense to say that that a tree risked its livelihood by growing in a particular place. That starts to sound like anthropomorphism. So there is an important part of the grammar of risk that picks out actions that are performed by intentional agents.

I am suggesting that, in the spirit of the kind of heuristics work that has been done by Tversky and Kahnemann and Slovic, we should understand this other sense of risk, what I call the act classifying the sense of risk, as a kind of heuristic. When we use the word risk in these contexts, we are picking out a class of actions. We are picking out a class of things that either people or organizations do. Under this definition, risks are actions that call for some sort of special consideration.

Next I want to discuss heuristics as a kind of cognitive filtering. When we call something a risk, we are saying that this deserves more consideration. We need to give it some thought. We need to do something with respect to it. And when we do not call something a risk, when we do not call it risky, we just go ahead and do it. These would be fairly routine, ordinary, habitual things that pass through the cognitive filter without detection. This cognitive filter may be culturally based or psychologically based. It is a way of telling us when to dedicate more resources, in the sense of time, energy, intellectual activity, or (socially) in terms of money to obtain information, write reports, or have committee meetings. It is a filter that tells us when it is important to do that and when it is not important to do that, because we tend to rely on habit, routine, or ordinary activities. There is a link between the intentionality and the cognitive filtering function because at least historically, but maybe not anymore, there has been very little point to devoting special attention to things that we cannot do anything about. So we look at actions that, if we did something else, then things would be different, or if I did something else, I might avoid a certain type of harm. We do not lump generic natural hazards, earthquakes, floods, tornadoes, and so on into that "could have acted otherwise" category.

So there is a sense in which, in this way of thinking about risk, things such as freak accidents and acts of God—and as well a background of hazards that characterize all of our daily activities—are not considered to be risks. Clearly accidents have some probability of harm associated with them, but they are not picked out by the cognitive filter that is associated with the word risk in an ordinary context.

I want to make a final point. Many times when people say that there is no risk associated with something, scientists interpret that as meaning that there is zero probability of harm. However, few people believe that there is zero probability of harm associated with any activity. But what is going on is that when someone makes a claim that "there is no risk," they are saying that it is something that has not made it through their cognitive filter. It is something that we do not devote any special attention to. We just keep doing what we have always been doing.

So there is a tension that arises between the way that the scientific risk assessment scientists talk about risk and this other notion of risk that is still very much alive in public discourse. Note that intention is irrelevant to the probability and harm conception of risk. Yet it is highly relevant to the cognitive filtering sense of risk.

When we start out with the event-predicting sense of risks, we are already involved in a process of deliberative optimizing. We want to know the probabilities and the level of harm because we are at least, at some level, making a risk-benefit trade-off decision. By deliberative, I mean that we are consciously thinking about options, we are consciously making a comparison, and we are, at least to some degree, consciously applying a decision rule about which way to go. We are doing very little consciously at the heuristics or the cognitive filtering level. This is the type of thing that happens before something even emerges in our world view as significant.

For the responses to act-classifying risks, there are three strategies that people follow, both individually and collectively, when they have decided that there is a risk in this broad sense of actions that call for special consideration.

The first is to eliminate the perceived source of risk to simplify one's life by saying "I don't even want to think about it. Just don't do it." A second thing someone will do is solve the problem of accountability. Who is going to be responsible in this particular situation? Am I responsible as the risk bearer? Are you responsible as the risk imposer? And if we get that satisfied satisfactorily, that may be the end of the story. We may not have done any work to either quantify or even approximate or estimate probabilities and consequences before we arrive at either of those two solutions. The third thing that we can do in this situation is to undertake a deliberation, to go to the trouble of trying to explicitly articulate—perhaps qualified, perhaps not—but explicitly articulate the dimensions of probability and harm and go through the process of making a deliberate conscious decision. This may be an individual working through a thought process or a group working through a social process. There is a sense in which what is going on in terms of a lot of the public debate is that the risk assessment community, and justifiably so, is already well into the process of deliberation. And the public is still sorting things out and talking about this as being risky in the sense that this is something that calls for a greater look and more care. And it is not clear that the public wants to resolve this problem by a deliberative strategy. They may be more receptive to resolving it by laying down some strict criteria of accountability or by simply eliminating the option from consideration.

What is the rationality that is implicit in this? Basically it would be quite irrational to engage in deliberative optimization with regard to all the potential choices that we face. If we did that, we would be spending all our time calculating probabilities and benefits and making comparative decisions. One after another there are hundreds of thousands of potential choices that we make every day, and it would be a tremendous waste of our cognitive resources to make deliberative decisions about all of them.

It is clear that there have to be some of these substitute rules that apportion deliberative resources and tell us when we are going to go though the explicit risk comparison. I am suggesting that although there is a clear sense in which deliberative optimizing gives us a very strong characterization of what would be rational behavior in a particular case, we need some type of heuristic operating in the background. This heuristic gives some sense of when it is the right time to get more information, when it is the right time to get a detailed risk assessment or risk calculation.

In looking at genetically engineered foods, I will assume that they score low on the probability and harm levels. That has been the scientific consensus, at least, although that consensus goes back and forth over time. Nevertheless, compared with microbial hazards, genetic engineering is not a serious risk issue with respect to the probability of harm. Compared with risks of global climate change, it is probably not even a serious environmental risk issue. Genetically modified food is not going to score very high on the two parameters of probability and degree of harm.

However, if we look at questions such as "Is it an action that is being undertaken intentionally?," it scores very high. It is not only an intentional (or deliberate) action, but it is very clearly promoted by the people that are undertaking the action as something that is new. The novelty of this activity is, in fact, a big element in the way it has been discussed. How does information on genetic engineering come to people? It often comes to them through channels that are perceived as strategic, meaning that it is through advertising or channels in which people with different points of view are debating one another over issues such as food safety policy or trade issues. Therefore, it is quite rational that it would tend to filter into a relatively high-risk category with respect to both the classifying and the information reliability.

Many people who are concerned about genetically modified organisms see it as an easily eliminable source of risk; they do not understand that there would be important costs associated with foregoing genetically engineered food altogether. Because of this, there has been a tendency to gravitate rather quickly toward the elimination strategy, at least in the minds of many people, and I do not believe that this is an irrational move for people to make. When the science and business communities strive to counter that move, they are perceived as engaging in strategic discourse. This cycle of factors tends to reinforce itself. In some respects, science institutions remain in a self-reinforcing cycle of increasing public skepticism about genetic engineering.

REFERENCES

Habermas, J. 1990. Discourse Ethics: Notes on a Program of Philosophical Justification, in The Communicative Ethics Debate, S. Behabib, and F. Dallmayer, eds. Cambridge, MA: Massachusetts Institute of Technology Press.

Kahneman, D., P. Slovic, and A. Tversky, eds. 1982. Judgment Under Uncertainty: Heuristics And Biases. New York: Cambridge University Press.

MacLean, D. 1990. Comparing values in environmental policies: moral issues and moral arguments. P.B. Hammond and R. Coppock, eds. Pp. 83–106 in Valuing Health Risks, Costs and Benefits for Environmental Decision Making. Washington, D.C.: National Acadamy Press.

Sagoff, M. 1985. Risk Benefit Analysis in Decisions Concerning Public Safety and Health, Dubuque, IA: Kendall/Hunt.

Shrader-Frechette, K. 1991. Risk and Rationality. Berkeley, California: University of California Press.

Slovic. P. 1987. Perception of Risk, Science 236:280–285.

Thompson, P.B. 1987. Agricultural Biotechnology and the Rhetoric of Risk: Some Conceptual Issues, The Environmental Professional, 9:316–326.

Thompson, P.B. 1991. Risk: Ethical Issues and Values, in Agricultural Biotechnology, Food Safety and Nutritional Quality for the Consumer, J.F. MacDonald, ed. National Agricultural Biotechnology Council (NABC) Report 2, Ithaca, N.Y.: NABC. Pp. 204–217.

Thompson, P.B. 1995. Risk and Responsibilities in Modern Agriculture, in Issues in Agricultural Bioethics, T.B. Mepham, G.A. Tucker, and J. Wiseman, eds. Nottingham: Nottingham University Press. Pp. 31–45.

Thompson, P.B. 1997a. Food Biotechnology in Ethical Perspective. London: Chapman and Hall.

Thompson, P.B. 1997b. Science Policy and Moral Purity: The Case of Animal Biotechnology, Agriculture and Human Values 14(1997):11–27.

Thompson, P.B. 1999. The Ethics of Truth-Telling and the Problem of Risk, Science and Engineering Ethics 5(4):489–511.

Thompson P.B. and W.E. Dean. 1996. Competing Conceptions of Risk, Risk: Health, Safety and Environment 7(4):361–384.

Wachbroit, R. 1991. Describing Risk, M.A. Levin and H.S. Strauss, eds., Risk Assessment in Genetic Engineering, New York: McGraw-Hill. Pp. 368–377.

Appendixes

.

Appendix A

Agreement on the Application of Sanitary and Phytosanitary Measures

Members,

Reaffirming that no Member should be prevented from adopting or enforcing measures necessary to protect human, animal or plant life or health, subject to the requirement that these measures are not applied in a manner which would constitute a means of arbitrary or unjustifiable discrimination between Members where the same conditions prevail or a disguised restriction on international trade;

Desiring to improve the human health, animal health and phytosanitary situation in all Members;

Noting that sanitary and phytosanitary measures are often applied on the basis of bilateral agreements or protocols;

Desiring the establishment of a multilateral framework of rules and disciplines to guide the development, adoption and enforcement of sanitary and phytosanitary measures in order to minimize their negative effects on trade;

Recognizing the important contribution that international standards, guidelines and recommendations can make in this regard;

Desiring to further the use of harmonized sanitary and phytosanitary measures between Members, on the basis of international standards, guidelines and recommendations developed by the relevant international organizations, including the Codex Alimentarius Commission, the International Office of

Epizootics, and the relevant international and regional organizations operating within the framework of the International Plant Protection Convention, without requiring Members to change their appropriate level of protection of human, animal or plant life or health;

Recognizing that developing country Members may encounter special difficulties in complying with the sanitary or phytosanitary measures of importing Members, and as a consequence in access to markets, and also in the formulation and application of sanitary or phytosanitary measures in their own territories, and desiring to assist them in their endeavours in this regard;

Desiring therefore to elaborate rules for the application of the provisions of GATT 1994 which relate to the use of sanitary or phytosanitary measures, in particular the provisions of Article XX(b)[1]

Hereby agree as follows:

Article 1: General Provisions

1. This Agreement applies to all sanitary and phytosanitary measures which may, directly or indirectly, affect international trade. Such measures shall be developed and applied in accordance with the provisions of this Agreement.

2. For the purposes of this Agreement, the definitions provided in Annex A shall apply.

3. The annexes are an integral part of this Agreement.

4. Nothing in this Agreement shall affect the rights of Members under the Agreement on Technical Barriers to Trade with respect to measures not within the scope of this Agreement.

Article 2: Basic Rights and Obligations

1. Members have the right to take sanitary and phytosanitary measures necessary for the protection of human, animal or plant life or health, provided that such measures are not inconsistent with the provisions of this Agreement.

2. Members shall ensure that any sanitary or phytosanitary measure is applied only to the extent necessary to protect human, animal or plant life or health, is based on scientific principles and is not maintained without sufficient scientific evidence, except as provided for in paragraph 7 of Article 5.

3. Members shall ensure that their sanitary and phytosanitary measures do not arbitrarily or unjustifiably discriminate between Members where identical or similar conditions prevail, including between their own territory and that of other Members. Sanitary and phytosanitary measures shall not be applied in a manner which would constitute a disguised restriction on international trade.

4. Sanitary or phytosanitary measures which conform to the relevant provisions of this Agreement shall be presumed to be in accordance with the

---

[1] In this Agreement, reference to Article XX(b) includes also the chapeau of that Article.

obligations of the Members under the provisions of GATT 1994 which relate to the use of sanitary or phytosanitary measures, in particular the provisions of Article XX(b).

Article 3: Harmonization

1. To harmonize sanitary and phytosanitary measures on as wide a basis as possible, Members shall base their sanitary or phytosanitary measures on international standards, guidelines or recommendations, where they exist, except as otherwise provided for in this Agreement, and in particular in paragraph 3.

2. Sanitary or phytosanitary measures which conform to international standards, guidelines or recommendations shall be deemed to be necessary to protect human, animal or plant life or health, and presumed to be consistent with the relevant provisions of this Agreement and of GATT 1994.

3. Members may introduce or maintain sanitary or phytosanitary measures which result in a higher level of sanitary or phytosanitary protection than would be achieved by measures based on the relevant international standards, guidelines or recommendations, if there is a scientific justification, or as a consequence of the level of sanitary or phytosanitary protection a Member determines to be appropriate in accordance with the relevant provisions of paragraphs 1 through 8 of Article 5[2] Notwithstanding the above, all measures which result in a level of sanitary or phytosanitary protection different from that which would be achieved by measures based on international standards, guidelines or recommendations shall not be inconsistent with any other provision of this Agreement.

4. Members shall play a full part, within the limits of their resources, in the relevant international organizations and their subsidiary bodies, in particular the Codex Alimentarius Commission, the International Office of Epizootics, and the international and regional organizations operating within the framework of the International Plant Protection Convention, to promote within these organizations the development and periodic review of standards, guidelines and recommendations with respect to all aspects of sanitary and phytosanitary measures.

5. The Committee on Sanitary and Phytosanitary Measures provided for in paragraphs 1 and 4 of Article 12 (referred to in this Agreement as the "Committee") shall develop a procedure to monitor the process of international harmonization and coordinate efforts in this regard with the relevant international organizations.

---

[2]For the purposes of paragraph 3 of Article 3, there is a scientific justification if, on the basis of an examination and evaluation of available scientific information in conformity with the relevant provisions of this Agreement, a Member determines that the relevant international standards, guidelines or recommendations are not sufficient to achieve its appropriate level of sanitary or phytosanitary protection.

Article 4: Equivalence

1. Members shall accept the sanitary or phytosanitary measures of other Members as equivalent, even if these measures differ from their own or from those used by other Members trading in the same product, if the exporting Member objectively demonstrates to the importing Member that its measures achieve the importing Member's appropriate level of sanitary or phytosanitary protection. For this purpose, reasonable access shall be given, upon request, to the importing Member for inspection, testing and other relevant procedures.

2. Members shall, upon request, enter into consultations with the aim of achieving bilateral and multilateral agreements on recognition of the equivalence of specified sanitary or phytosanitary measures.

Article 5: Assessment of Risk and Determination of the Appropriate Level of Sanitary or Phytosanitary Protection

1. Members shall ensure that their sanitary or phytosanitary measures are based on an assessment, as appropriate to the circumstances, of the risks to human, animal or plant life or health, taking into account risk assessment techniques developed by the relevant international organizations.

2. In the assessment of risks, Members shall take into account available scientific evidence; relevant processes and production methods; relevant inspection, sampling and testing methods; prevalence of specific diseases or pests; existence of pest- or disease-free areas; relevant ecological and environmental conditions; and quarantine or other treatment.

3. In assessing the risk to animal or plant life or health and determining the measure to be applied for achieving the appropriate level of sanitary or phytosanitary protection from such risk, Members shall take into account as relevant economic factors: the potential damage in terms of loss of production or sales in the event of the entry, establishment or spread of a pest or disease; the costs of control or eradication in the territory of the importing Member; and the relative cost-effectiveness of alternative approaches to limiting risks.

4. Members should, when determining the appropriate level of sanitary or phytosanitary protection, take into account the objective of minimizing negative trade effects.

5. With the objective of achieving consistency in the application of the concept of appropriate level of sanitary or phytosanitary protection against risks to human life or health, or to animal and plant life or health, each Member shall avoid arbitrary or unjustifiable distinctions in the levels it considers to be appropriate in different situations, if such distinctions result in discrimination or a disguised restriction on international trade. Members shall cooperate in the Committee, in accordance with paragraphs 1, 2 and 3 of Article 12, to develop guidelines to further the practical implementation of this provision. In developing the guidelines, the Committee shall take into account all relevant factors, including the exceptional character of human health risks to which people voluntarily expose themselves.

6. Without prejudice to paragraph 2 of Article 3, when establishing or maintaining sanitary or phytosanitary measures to achieve the appropriate level of sanitary or phytosanitary protection, Members shall ensure that such measures are not more trade-restrictive than required to achieve their appropriate level of sanitary or phytosanitary protection, taking into account technical and economic feasibility.[3]

7. In cases where relevant scientific evidence is insufficient, a Member may provisionally adopt sanitary or phytosanitary measures on the basis of available pertinent information, including that from the relevant international organizations as well as from sanitary or phytosanitary measures applied by other Members. In such circumstances, Members shall seek to obtain the additional information necessary for a more objective assessment of risk and review the sanitary or phytosanitary measure accordingly within a reasonable period of time.

8. When a Member has reason to believe that a specific sanitary or phytosanitary measure introduced or maintained by another Member is constraining, or has the potential to constrain, its exports and the measure is not based on the relevant international standards, guidelines or recommendations, or such standards, guidelines or recommendations do not exist, an explanation of the reasons for such sanitary or phytosanitary measure may be requested and shall be provided by the Member maintaining the measure.

Article 6: Adaptation to Regional Conditions, Including Pest- or Disease-Free Areas and Areas of Low Pest or Disease Prevalence

1. Members shall ensure that their sanitary or phytosanitary measures are adapted to the sanitary or phytosanitary characteristics of the area - whether all of a country, part of a country, or all or parts of several countries - from which the product originated and to which the product is destined. In assessing the sanitary or phytosanitary characteristics of a region, Members shall take into account, inter alia, the level of prevalence of specific diseases or pests, the existence of eradication or control programmes, and appropriate criteria or guidelines which may be developed by the relevant international organizations.

2. Members shall, in particular, recognize the concepts of pest- or disease-free areas and areas of low pest or disease prevalence. Determination of such areas shall be based on factors such as geography, ecosystems, epidemiological surveillance, and the effectiveness of sanitary or phytosanitary controls.

3. Exporting Members claiming that areas within their territories are pest- or disease-free areas or areas of low pest or disease prevalence shall provide the necessary evidence thereof in order to objectively demonstrate to the importing Member that such areas are, and are likely to remain, pest- or disease-free areas

---

[3]For purposes of paragraph 6 of Article 5, a measure is not more trade-restrictive than required unless there is another measure, reasonably available taking into account technical and economic feasibility, that achieves the appropriate level of sanitary or phytosanitary protection and is significantly less restrictive to trade.

or areas of low pest or disease prevalence, respectively. For this purpose, reasonable access shall be given, upon request, to the importing Member for inspection, testing and other relevant procedures.

Article 7: Transparency

Members shall notify changes in their sanitary or phytosanitary measures and shall provide information on their sanitary or phytosanitary measures in accordance with the provisions of Annex B.

Article 8: Control, Inspection and Approval Procedures

Members shall observe the provisions of Annex C in the operation of control, inspection and approval procedures, including national systems for approving the use of additives or for establishing tolerances for contaminants in foods, beverages or feedstuffs, and otherwise ensure that their procedures are not inconsistent with the provisions of this Agreement.

Article 9: Technical Assistance

1. Members agree to facilitate the provision of technical assistance to other Members, especially developing country Members, either bilaterally or through the appropriate international organizations. Such assistance may be, inter alia, in the areas of processing technologies, research and infrastructure, including in the establishment of national regulatory bodies, and may take the form of advice, credits, donations and grants, including for the purpose of seeking technical expertise, training and equipment to allow such countries to adjust to, and comply with, sanitary or phytosanitary measures necessary to achieve the appropriate level of sanitary or phytosanitary protection in their export markets.

2. Where substantial investments are required in order for an exporting developing country Member to fulfil the sanitary or phytosanitary requirements of an importing Member, the latter shall consider providing such technical assistance as will permit the developing country Member to maintain and expand its market access opportunities for the product involved.

Article 10: Special and Differential Treatment

1. In the preparation and application of sanitary or phytosanitary measures, Members shall take account of the special needs of developing country Members, and in particular of the least-developed country Members.

2. Where the appropriate level of sanitary or phytosanitary protection allows scope for the phased introduction of new sanitary or phytosanitary measures, longer time-frames for compliance should be accorded on products of

interest to developing country Members so as to maintain opportunities for their exports.

3. With a view to ensuring that developing country Members are able to comply with the provisions of this Agreement, the Committee is enabled to grant to such countries, upon request, specified, time-limited exceptions in whole or in part from obligations under this Agreement, taking into account their financial, trade and development needs.

4. Members should encourage and facilitate the active participation of developing country Members in the relevant international organizations.

Article 11: Consultations and Dispute Settlement

1. The provisions of Articles XXII and XXIII of GATT 1994 as elaborated and applied by the Dispute Settlement Understanding shall apply to consultations and the settlement of disputes under this Agreement, except as otherwise specifically provided herein.

2. In a dispute under this Agreement involving scientific or technical issues, a panel should seek advice from experts chosen by the panel in consultation with the parties to the dispute. To this end, the panel may, when it deems it appropriate, establish an advisory technical experts group, or consult the relevant international organizations, at the request of either party to the dispute or on its own initiative.

3. Nothing in this Agreement shall impair the rights of Members under other international agreements, including the right to resort to the good offices or dispute settlement mechanisms of other international organizations or established under any international agreement.

Article 12: Administration

1. A Committee on Sanitary and Phytosanitary Measures is hereby established to provide a regular forum for consultations. It shall carry out the functions necessary to implement the provisions of this Agreement and the furtherance of its objectives, in particular with respect to harmonization. The Committee shall reach its decisions by consensus.

2. The Committee shall encourage and facilitate ad hoc consultations or negotiations among Members on specific sanitary or phytosanitary issues. The Committee shall encourage the use of international standards, guidelines or recommendations by all Members and, in this regard, shall sponsor technical consultation and study with the objective of increasing coordination and integration between international and national systems and approaches for approving the use of food additives or for establishing tolerances for contaminants in foods, beverages or feedstuffs.

3. The Committee shall maintain close contact with the relevant international organizations in the field of sanitary and phytosanitary protection, especially with the Codex Alimentarius Commission, the International Office of

Epizootics, and the Secretariat of the International Plant Protection Convention, with the objective of securing the best available scientific and technical advice for the administration of this Agreement and in order to ensure that unnecessary duplication of effort is avoided.

4. The Committee shall develop a procedure to monitor the process of international harmonization and the use of international standards, guidelines or recommendations. For this purpose, the Committee should, in conjunction with the relevant international organizations, establish a list of international standards, guidelines or recommendations relating to sanitary or phytosanitary measures which the Committee determines to have a major trade impact. The list should include an indication by Members of those international standards, guidelines or recommendations which they apply as conditions for import or on the basis of which imported products conforming to these standards can enjoy access to their markets. For those cases in which a Member does not apply an international standard, guideline or recommendation as a condition for import, the Member should provide an indication of the reason therefor, and, in particular, whether it considers that the standard is not stringent enough to provide the appropriate level of sanitary or phytosanitary protection. If a Member revises its position, following its indication of the use of a standard, guideline or recommendation as a condition for import, it should provide an explanation for its change and so inform the Secretariat as well as the relevant international organizations, unless such notification and explanation is given according to the procedures of Annex B.

5. In order to avoid unnecessary duplication, the Committee may decide, as appropriate, to use the information generated by the procedures, particularly for notification, which are in operation in the relevant international organizations.

6. The Committee may, on the basis of an initiative from one of the Members, through appropriate channels invite the relevant international organizations or their subsidiary bodies to examine specific matters with respect to a particular standard, guideline or recommendation, including the basis of explanations for non-use given according to paragraph 4.

7. The Committee shall review the operation and implementation of this Agreement three years after the date of entry into force of the WTO Agreement, and thereafter as the need arises. Where appropriate, the Committee may submit to the Council for Trade in Goods proposals to amend the text of this Agreement having regard, inter alia, to the experience gained in its implementation.

Article 13: Implementation

Members are fully responsible under this Agreement for the observance of all obligations set forth herein. Members shall formulate and implement positive measures and mechanisms in support of the observance of the provisions of this Agreement by other than central government bodies. Members shall take such reasonable measures as may be available to them to ensure that non-governmental entities within their territories, as well as regional bodies in which relevant entities within their territories are members, comply with the relevant provisions of this Agreement. In addition, Members shall not take measures which have the effect of, directly or indirectly, requiring or encouraging such regional or non-governmental entities, or local governmental bodies, to act in a manner inconsistent with the provisions of this Agreement. Members shall ensure that they rely on the services of non-governmental entities for implementing sanitary or phytosanitary measures only if these entities comply with the provisions of this Agreement.

Article 14: Final Provisions

The least-developed country Members may delay application of the provisions of this Agreement for a period of five years following the date of entry into force of the WTO Agreement with respect to their sanitary or phytosanitary measures affecting importation or imported products. Other developing country Members may delay application of the provisions of this Agreement, other than paragraph 8 of Article 5 and Article 7, for two years following the date of entry into force of the WTO Agreement with respect to their existing sanitary or phytosanitary measures affecting importation or imported products, where such application is prevented by a lack of technical expertise, technical infrastructure or resources.

ANNEX A: DEFINITIONS[4]

1. *Sanitary or phytosanitary measure.* Any measure applied:

(a) to protect animal or plant life or health within the territory of the Member from risks arising from the entry, establishment or spread of pests, diseases, disease-carrying organisms or disease-causing organisms;
(b) to protect human or animal life or health within the territory of the Member from risks arising from additives, contaminants, toxins or disease-causing organisms in foods, beverages or feedstuffs;
(c) to protect human life or health within the territory of the Member from risks arising from diseases carried by animals, plants or products thereof, or from the entry, establishment or spread of pests; or
(d) to prevent or limit other damage within the territory of the Member from the entry, establishment or spread of pests.

Sanitary or phytosanitary measures include all relevant laws, decrees, regulations, requirements and procedures including, inter alia, end product criteria; processes and production methods; testing, inspection, certification and approval procedures; quarantine treatments including relevant requirements

---

[4]For the purpose of these definitions, "animal" includes fish and wild fauna; "plant" includes forests and wild flora; "pests" include weeds; and "contaminants" include pesticide and veterinary drug residues and extraneous matter.

associated with the transport of animals or plants, or with the materials necessary for their survival during transport; provisions on relevant statistical methods, sampling procedures and methods of risk assessment; and packaging and labelling requirements directly related to food safety.

2. *Harmonization.* The establishment, recognition and application of common sanitary and phytosanitary measures by different Members.

3. *International standards, guidelines and recommendations*

(a) for food safety, the standards, guidelines and recommendations established by the Codex Alimentarius Commission relating to food additives, veterinary drug and pesticide residues, contaminants, methods of analysis and sampling, and codes and guidelines of hygienic practice;

(b) for animal health and zoonoses, the standards, guidelines and recommendations developed under the auspices of the International Office of Epizootics;

(c) for plant health, the international standards, guidelines and recommendations developed under the auspices of the Secretariat of the International Plant Protection Convention in cooperation with regional organizations operating within the framework of the International Plant Protection Convention; and

(d) for matters not covered by the above organizations, appropriate standards, guidelines and recommendations promulgated by other relevant international organizations open for membership to all Members, as identified by the Committee.

4. *Risk assessment.* The evaluation of the likelihood of entry, establishment or spread of a pest or disease within the territory of an importing Member according to the sanitary or phytosanitary measures which might be applied, and of the associated potential biological and economic consequences; or the evaluation of the potential for adverse effects on human or animal health arising from the presence of additives, contaminants, toxins or disease-causing organisms in food, beverages or feedstuffs.

5. *Appropriate level of sanitary or phytosanitary protection.* The level of protection deemed appropriate by the Member establishing a sanitary or phytosanitary measure to protect human, animal or plant life or health within its territory.

NOTE: Many Members otherwise refer to this concept as the "acceptable level of risk".

6. *Pest- or disease-free area.* An area, whether all of a country, part of a country, or all or parts of several countries, as identified by the competent authorities, in which a specific pest or disease does not occur.

NOTE: A pest- or disease-free area may surround, be surrounded by, or be adjacent to an area - whether within part of a country or in a geographic region which includes parts of or all of several countries -in which a specific pest or disease is known to occur but is subject to regional control measures such as the establishment of protection, surveillance and buffer zones which will confine or eradicate the pest or disease in question.

7. *Area of low pest or disease prevalence.* An area, whether all of a country, part of a country, or all or parts of several countries, as identified by the competent authorities, in which a specific pest or disease occurs at low levels and which is subject to effective surveillance, control or eradication measures.

ANNEX B: TRANSPARENCY OF SANITARY AND PHYTOSANITARY REGULATIONS

Publication of Regulations

1. Members shall ensure that all sanitary and phytosanitary regulations[5] which have been adopted are published promptly in such a manner as to enable interested Members to become acquainted with them.

2. Except in urgent circumstances, Members shall allow a reasonable interval between the publication of a sanitary or phytosanitary regulation and its entry into force in order to allow time for producers in exporting Members, and particularly in developing country Members, to adapt their products and methods of production to the requirements of the importing Member.

Enquiry Points

3. Each Member shall ensure that one enquiry point exists which is responsible for the provision of answers to all reasonable questions from interested Members as well as for the provision of relevant documents regarding:

(a) any sanitary or phytosanitary regulations adopted or proposed within its territory;

(b) any control and inspection procedures, production and quarantine treatment, pesticide tolerance and food additive approval procedures, which are operated within its territory;

(c) risk assessment procedures, factors taken into consideration, as well as the determination of the appropriate level of sanitary or phytosanitary protection;

(d) the membership and participation of the Member, or of relevant bodies within its territory, in international and regional sanitary and phytosanitary organizations and systems, as well as in bilateral and multilateral agreements and arrangements within the scope of this Agreement, and the texts of such agreements and arrangements.

---

[5]Sanitary and phytosanitary measures such as laws, decrees or ordinances which are applicable generally.

4. Members shall ensure that where copies of documents are requested by interested Members, they are supplied at the same price (if any), apart from the cost of delivery, as to the nationals[6] of the Member concerned.

Notification Procedures

5. Whenever an international standard, guideline or recommendation does not exist or the content of a proposed sanitary or phytosanitary regulation is not substantially the same as the content of an international standard, guideline or recommendation, and if the regulation may have a significant effect on trade of other Members, Members shall:

(a) publish a notice at an early stage in such a manner as to enable interested Members to become acquainted with the proposal to introduce a particular regulation;

(b) notify other Members, through the Secretariat, of the products to be covered by the regulation together with a brief indication of the objective and rationale of the proposed regulation. Such notifications shall take place at an early stage, when amendments can still be introduced and comments taken into account;

(c) provide upon request to other Members copies of the proposed regulation and, whenever possible, identify the parts which in substance deviate from international standards, guidelines or recommendations;

(d) without discrimination, allow reasonable time for other Members to make comments in writing, discuss these comments upon request, and take the comments and the results of the discussions into account.

6. However, where urgent problems of health protection arise or threaten to arise for a Member, that Member may omit such of the steps enumerated in paragraph 5 of this Annex as it finds necessary, provided that the Member:

(a) immediately notifies other Members, through the Secretariat, of the particular regulation and the products covered, with a brief indication of the objective and the rationale of the regulation, including the nature of the urgent problem(s);

(b) provides, upon request, copies of the regulation to other Members;

(c) allows other Members to make comments in writing, discusses these comments upon request, and takes the comments and the results of the discussions into account.

7. Notifications to the Secretariat shall be in English, French or Spanish.

---

[6]When "nationals" are referred to in this Agreement, the term shall be deemed, in the case of a separate customs territory Member of the WTO, to mean persons, natural or legal, who are domiciled or who have a real and effective industrial or commercial establishment in that customs territory.

8. Developed country Members shall, if requested by other Members, provide copies of the documents or, in case of voluminous documents, summaries of the documents covered by a specific notification in English, French or Spanish.

9. The Secretariat shall promptly circulate copies of the notification to all Members and interested international organizations and draw the attention of developing country Members to any notifications relating to products of particular interest to them.

10. Members shall designate a single central government authority as responsible for the implementation, on the national level, of the provisions concerning notification procedures according to paragraphs 5, 6, 7 and 8 of this Annex.

General Reservations

11. Nothing in this Agreement shall be construed as requiring:

(a) the provision of particulars or copies of drafts or the publication of texts other than in the language of the Member except as stated in paragraph 8 of this Annex; or

(b) Members to disclose confidential information which would impede enforcement of sanitary or phytosanitary legislation or which would prejudice the legitimate commercial interests of particular enterprises.

ANNEX C: CONTROL, INSPECTION AND APPROVAL PROCEDURES[7]

1. Members shall ensure, with respect to any procedure to check and ensure the fulfilment of sanitary or phytosanitary measures, that:

(a) such procedures are undertaken and completed without undue delay and in no less favourable manner for imported products than for like domestic products;

(b) the standard processing period of each procedure is published or that the anticipated processing period is communicated to the applicant upon request; when receiving an application, the competent body promptly examines the completeness of the documentation and informs the applicant in a precise and complete manner of all deficiencies; the competent body transmits as soon as possible the results of the procedure in a precise and complete manner to the applicant so that corrective action may be taken if necessary; even when the application has deficiencies, the competent body proceeds as far as practicable with the procedure if the applicant so requests; and that upon request, the

---

[7]Control, inspection and approval procedures include, inter alia, procedures for sampling, testing and certification.

applicant is informed of the stage of the procedure, with any delay being explained;

(c) information requirements are limited to what is necessary for appropriate control, inspection and approval procedures, including for approval of the use of additives or for the establishment of tolerances for contaminants in food, beverages or feedstuffs;

(d) the confidentiality of information about imported products arising from or supplied in connection with control, inspection and approval is respected in a way no less favourable than for domestic products and in such a manner that legitimate commercial interests are protected;

(e) any requirements for control, inspection and approval of individual specimens of a product are limited to what is reasonable and necessary;

(f) any fees imposed for the procedures on imported products are equitable in relation to any fees charged on like domestic products or products originating in any other Member and should be no higher than the actual cost of the service;

(g) the same criteria should be used in the siting of facilities used in the procedures and the selection of samples of imported products as for domestic products so as to minimize the inconvenience to applicants, importers, exporters or their agents;

(h) whenever specifications of a product are changed subsequent to its control and inspection in light of the applicable regulations, the procedure for the modified product is limited to what is necessary to determine whether adequate confidence exists that the product still meets the regulations concerned; and

(i) a procedure exists to review complaints concerning the operation of such procedures and to take corrective action when a complaint is justified.

Where an importing Member operates a system for the approval of the use of food additives or for the establishment of tolerances for contaminants in food, beverages or feedstuffs which prohibits or restricts access to its domestic markets for products based on the absence of an approval, the importing Member shall consider the use of a relevant international standard as the basis for access until a final determination is made.

2. Where a sanitary or phytosanitary measure specifies control at the level of production, the Member in whose territory the production takes place shall provide the necessary assistance to facilitate such control and the work of the controlling authorities.

3. Nothing in this Agreement shall prevent Members from carrying out reasonable inspection within their own territories.

Source: World Trade Organization. Available on-line at:
<http://www.wto.org/wto/goods/spsagr.htm>. March 9, 2000.

Appendix B

Conference Program

INCORPORATING SCIENCE, ECONOMICS, SOCIOLOGY AND POLITICS IN SANITARY AND PHYTOSANITARY STANDARDS IN INTERNATIONAL TRADE

January 25–27, 1999
National Academy of Sciences and Engineering
Beckman Center, Irvine, California

JANUARY 25, 1999

7:15–8:00 p.m. Keynote Address
Historical and Social Science Perspectives on the Role of Risk Assessment and Science in Protecting the Domestic Economy: Some Background
G. Edward Schuh, Humphrey Institute of Public Affairs, University of Minnesota

JANUARY 26, 1999

8:20–8:30 a.m. Introduction
                V. Kerry Smith, North Carolina State University, Department of
                Agriculture and Resource Economics

Session I: **Agricultural Trade, Risk Assessment, and the Role of**
                Culture in Risk Management
                Moderators: *Raymond A. Jussaume, Jr., Department of Rural*
                Sociology, Washington State University
                Peter Kareiva, Department of Zoology, University of
                Washington

8:30–9:00   Overview of SPS and Agricultural Trade
                Donna Roberts, Economic Research Service, USDA

9:00–9:15   Discussion

9:15–9:45   An Overview of Risk Assessment
                John D. Stark, Department of Entomology, Washington State
                University

9:45–10:00   Discussion

10:00–10:30   BREAK

10:30–11:00   Technological Risk and Cultures of Rationality
                Sheila Jasanoff, John F. Kennedy School of Government,
                Harvard University

11:00–11:15   Discussion

Session II   **General Case Studies**
                Moderator: *Julie Caswell, Department of Resource Economics,*
                University of Massachusetts

CASE STUDY 1: MEAT SLAUGHTERING AND PROCESSING
PRACTICES (INCLUDES VETERINARY EQUIVALENCE AND HAACP)

11:15–11:45   *Bent Nielsen, Veterinary and Food Advisory Services,*
                Copenhagen, Denmark

11:45–12:15   *Bruce A. Silverglade, Center for Science in the Public Interest,*
                Washington, D.C.

12:15–1:30   BREAK

CASE STUDY 2: PLANT QUARANTINES AND HASS AVOCADOS

1:30–2:00   *Walther Enkerlin Hoeflich, Mexican Stone Fruit Inspection Program, Clovis, California*

2:00–2:30   *David Vogel, Haas School of Business, University of California, Berkeley*

CASE STUDY 3: GENETICALLY MODIFIED ORGANISMS

2:30–3:00   *Peter Kareiva, Department of Zoology, University of Washington*

3:00–3:30   *Paul B. Thompson, Department of Philosophy, Purdue University*

3:35–4:00   BREAK

Session III   **Case Study Discussions**

4:00–5:30   Breakout group discussions

5:30   ADJOURN

JANUARY 27, 1999

8:20–8:30 a.m.   Announcements
V. Kerry Smith, North Carolina State University, Department of Agriculture and Resource Economics

8:30–9:30   Individuals reports from breakout groups

Session IV   **Political And Ecological Economy**
Moderators: *V. Kerry Smith, North Carolina State University, Department of Agriculture and Resource Economics*
David Vogel, Haas School of Business, University of California, Berkeley

9:30–10:00   Ecological Impacts
Karen Goodell, Department of Ecology and Evolution, State University of New York at Stoney Brook
Peter Kareiva, Department of Zoology, University of Washington

10:00–10:45   Discussion

10:45–11:15   The Political Economy
              David G. Victor, Council on Foreign Relations, New York

11:15–11:30   Discussion

11:30–12:00   Accounting for Consumer Preferences in International Trade
              *Jean-Christophe Bureau, Station d'Economie et Sociologie
              Rurales, Institute for Agricultural Research, Grignon, France*

12:00–12:15   Discussion

12:15–1:30   BREAK

Session V   **Resolving Current SPS Trade Disputes and Establishing a
              Basis for Defusing Future Conflicts (see questions below)**
              Moderators: *Timothy Josling, Institute of International Studies,
              Stanford University*
              D. Warner North, NorthWorks, Inc., Belmont, Calif.

1:30–3:00   Panels and General Discussion
              *Linda Horton, International Policy, Food and Drug
              Administration*
              *Dan Sumner, Department of Agriculture and Resource
              Economics, University of California, Davis*
              *James H. McDonald, Division of Behavioral and Cultural
              Studies, University of Texas at San Antonio*
              *Julie Caswell, Department of Resource Economics, University
              of Massachusetts*

3:00–3:15   Closing Comments
              *V. Kerry Smith, North Carolina State University, Department of
              Agriculture and Resource Economics*

3:15   ADJOURN

DISCUSSION QUESTIONS

Session III: Case Study Discussions

Risk analyses incorporate scientific information to measure and describe impacts on health and the environment from exposure to contaminants. They require information of the substances, sources, exposure median and patterns, events at risk, affected populations and response options. Using the conference case studies as a basis for discussion, please provide your input to the following questions:

1. Who (e.g., nations, organizations) performed a risk assessment? Why? What was the outcome?
2. How was the risk assessment process managed? What sciences were involved and what was lacking in the analysis? How well were the natural and social sciences used? How did cultural values and beliefs influence the way that various countries/regions assessed the soundness of science in the process?
3. What was the source of the problem or solution that led to that outcome (e.g., regulatory structure)?
4. What sciences should be included and how can these sciences be integrated in risk analyses used for SPS decisionmaking?

Session V: Resolving Current SPS Trade Disputes and Establishing a Basis for Defusing Future Conflicts

1. What is the current role of natural and social sciences in SPS decisionmaking?
2. What is missing from the decision-making process?
3. What is the role of private sector standards and voluntary labeling systems? How far can the public authorities rely on the industry to regulate itself? Would such self-regulation work in global markets? Should one try to harmonize such liability laws across countries?
4. What public educational needs are there in this area? Should governments coordinate their educational efforts? Is there a role for international organizations in addressing consumer concerns directly? What research needs have been identified in the area? What institutional support might be warranted for this research?
5. What changes in the procedures of national regulatory agencies would assist with the prevention of trade conflicts? Should these agencies coordinate more, or is the responsibility for control of domestic market? Are these agencies independent of domestic vested interests, such as producer groups?
6. What more can be done to promote the use of international standards? Has the experience with *Codex*, IPPC, and the OIE been satisfactory in resolving or reducing trade frictions? What are the effective limits to the use of

harmonized standards? Is there a role for international agencies which would have the responsibility of setting standards rather than just suggesting them? What is the role of regional and bilateral SPS agreements? Is mutual recognition of national standards a viable option?

7. What is a reasonable conceptual and empirical framework for incorporating cultural and scientific factors in SPS decisionmaking?

Appendix C

Program Participants

JEAN-CHRISTOPHE BUREAU is an economist and research director with the French Institute for Agricultural Research in Thiverval-Grignon, France. He holds a Ph.D. in economics from University Paris-Sorbonne. His research focus is on agricultural productivity and consumer concerns with trade.

JULIE CASWELL is a professor at the Department of Resource Economics at the University of Massachusetts. Her research areas include food quality and safety, strategic decision making, industrial organization, risk assessment and benefit/cost analysis. Her research focuses on understanding the operation of domestic and international food systems, with particular interest in the economics of food quality, especially the quality attributes of safety and nutrition, and international trade. Caswell received her Ph.D. in agricultural economics and economics from the University of Wisconsin.

WALTHER ENKERLIN HOEFLICH is currently working as a Technical Officer for the International Atomic Energy Agency (IAEA) in Vienna, Austria. He worked for the Mexican Dirección General de Sanidad Vegetal, Secretaría de Agricultura, Ganadería y Desarrollo Rural (DGSV/SAGAR), in Sterile Insect Technique (SIT) based area-wide fruit fly control programs for 11 years. He has been a consultant for the FAO/IAEA joint division in SIT feasibility assessments, technology transfer and program implementation. Hoeflich holds a Ph.D. degree in applied entomology from the University of London.

KAREN GOODELL is a Ph.D. student and teaching assistant in the Department of Ecology and Evolution at the State University of New York at Stony Brook. The recipient of numerous grants, fellowships, and awards, Goodell's primary research focus is on the effects of introduced honeybees on native solitary bees.

SHEILA JASANOFF is professor of science and public policy at the John F. Kennedy School of Government and School of Public Health at Harvard University. A lawyer by training, her research career has been devoted to the interaction of law, science, and politics in democratic societies and she has pioneered the integration of perspectives from the social studies of science with legal and policy analysis. She holds a Ph.D. from Harvard University and a J.D. degree from Harvard Law School.

TIMOTHY JOSLING is an agricultural economist and professor in the Food Research Institute at Stanford University. Josling's research interests center on industrial country agricultural policies, international trade in agricultural products, and the process of economic integration. He is currently involved in studies of the reform of the agricultural trading system and agriculture trade policies in the World Trade Organization, North American Free Trade Agreement, and MERCOSUR, countries of the Caribbean Basin, and the European Union (EU). Josling received a Ph.D. in agricultural economics from Michigan State University.

RAYMOND A. JUSSAUME, JR., is a rural sociologist and currently holds a joint appointment in research, extension, and teaching at the Department of Rural Sociology at Washington State University. Jussaume's primary emphases include international agricultural marketing and trade; community and development studies; and he specializes in cross-cultural issues of markets, consumer preferences, and values for multiple crops and animal products. Jussaume received his Ph.D. in development sociology from Cornell University. He also has an M.A. in political science (with a certificate in global policy studies) from the University of Georgia.

PETER KAREIVA is an ecologist in the Department of Zoology at the University of Washington. Kareiva's research is focused on spatial heterogeneity influences on species interactions and building of ecological models that can be scaled up to address multispecies interactions and environmental issues. He also is examining ecological theory of biocontrol, population biology of herbivorous insects, impacts of nonindigenous organisms, and cross-fertilization between transgenic crops and wild relatives. Kareiva received his Ph.D. in ecology and evolution from Cornell University.

JAMES H. McDONALD is assistant professor of anthropology, University of Texas at San Antonio. His areas of specialization include economic and political/legal anthropology, political economy and agricultural development in Mexico, Latin America, and the United States. He holds a Ph.D. in anthropology from Arizona State University.

BENT NIELSEN is head of Section for Zoonotic Diseases, Veterinary and Food Safety Service at the Federation of Danish Pig Producers and Slaughterhouses in Copenhagen, Denmark. His research efforts have focused on food safety issues related to microbial contamination of meat and he has been instrumental in developing serological testing strategies for controlling *Salmonella* in swine herds and *Salmonella* contamination of pork. He holds a D.V.M. from The Royal Veterinary and Agricultural University in Denmark and a Ph.D. in veterinary microbiology and immunology from The Royal Veterinary and Agricultural University, Denmark.

D. WARNER NORTH is president and principal scientist of NorthWorks, Inc., a consulting firm in Belmont, California, and a consulting professor in the Department of Engineering-Economic Systems and Operations Research at Stanford University. He has participated in more than a dozen National Research Council studies on environmental risk, most recently *Understanding Risk: Informing Decisions in a Democratic Society* (1996). He has carried out a variety of applications of decision analysis and risk analysis, from quarantine policy for the exploration of Mars to management of toxic substances and nuclear waste. North received his Ph.D. in operations research from Stanford.

DONNA ROBERTS is an economist with the Economic Research Service of the U.S. Department of Agriculture. She is currently working at the U.S. Trade Representative's mission to the World Trade Organization in Geneva, Switzerland.

G. EDWARD SCHUH is the Orville and Jane Freeman Chair in International Trade and Investment Policy at the University of Minnesota. Prior to assuming the Freeman Chair, he was dean of the Humphrey Institute of Public Affairs for 10 years. Dr. Schuh's expertise includes international teaching, consulting, and advising experience in Latin America and India, and with the U.S. government. Schuh is a member of the National Research Council Board on Agriculture and Natural Resources.

BRUCE A. SILVERGLADE, Esq., is legal affairs director of the Center for Science in the Public Interest (CSPI) in Washington, D.C. CSPI is an international nonprofit consumer advocacy organization with approximately 1,000,000 members. Silverglade coordinates CSPI's advocacy activities in a variety of areas involving food and dietary supplement regulation, supervises court litigation involving consumer health issues, and leads CSPI's participation in matters before the *Codex Alimentarius* Commission. He received his law degree from Boston College and a B.A. in political science from the University of Illinois.

V. KERRY SMITH is director of Center for Environmental and Resource Economic Policy, and University Distinguished Professor at the Department of Agricultural and Resource Economics, North Carolina State University. His current research interests include the conceptual and empirical issues in valuing nonmarket environmental resources such as clean air and water, the use of computable general equilibrium models to understand how environment policies influence international trade, and public policies involving environmental risks. Smith received his Ph.D. in economics from Rutgers University.

JOHN D. STARK is an exocologist and entomologist at Washington State University. His research interests involve estimating the fate of pesticides and other xenobiotics in the environment and estimating their effects on populations, communities, and food webs, with particular emphasis on demographic toxicology and modeling. He holds a Ph.D. in pesticide toxicology and entomology from the University of Hawaii.

PAUL B. THOMPSON is a professor of philosophy at Purdue University. His focus is on ethics in research and public policy in diverse areas such as food biotechnology, environment, and agriculture. He has a Ph.D. in philosophy from the State University of New York at Stony Brook.

DAVID G. VICTOR currently is the Robert W. Johnson, Jr., Fellow for Science and Technology at the Council on Foreign Relations in New York. His research focuses on how science and technology affect U.S. foreign policy. Victor received his Ph.D. in political science at the Massachusetts Institute of Technology.

DAVID VOGEL is a political scientist and currently holds the George Quist Chair in Business Ethics at the University of California, Berkeley. His current research interests integrate environmental, consumer, and trade policy and explicitly challenges the conventional wisdom that trade liberalization and agreements to promote free trade invariably undermine national health, safety, and environmental standards. Vogel received his Ph.D. in political science from Princeton University. He is the author of *Trading Up: Consumer and Environmental Regulation in a Global Economy*, Harvard University Press, 1995.

Appendix D

Conference Participants

Nell Ahl, Office of Risk Assessment and Cost Benefit Analysis, U.S. Department of Agriculture, Washington, D.C.

Nicole Ballenger, Market and Trade Economics Division, Economic Research Service, U.S. Department of Agriculture, Washington, D.C.

Rebecca Bech, Animal and Plant Health Inspection Service, U.S. Department of Agriculture, Riverdale, Maryland

Beth Burrows, The Edmonds Institute, Edmonds, Washington, D.C.

Lawrence Busch, Department of Sociology, Michigan State University, East Lansing

Jean Buzby, Food and Rural Economics Division, Economic Research Service, U.S. Department of Agriculture, Washington, D.C.

Faith Campbell, American Lands Association, Washington, D.C.

Patrick Clerkin, Codex Office, U.S. Department of Agriculture, Washington, D.C.

Marsha Echols, Howard University School of Law, Washington, D.C.

Daniel Fieselmann, Animal and Plant Health Inspection Service, U.S. Department of Agriculture, Raleigh, North Carolina

Richard Fite, Risk Analysis Systems, Animal and Plant Health Inspection Service, U.S. Department of Agriculture, Riverdale, Maryland

Ken Forsythe, Veterinary Services Centers for Epidemiology and Animal Health, Animal and Plant Health Inspection Service, U.S. Department of Agriculture, Fort Collins, Colorado

William Friedland, Community Studies Department, University of California, Santa Cruz

Elise Golan, President's Council of Economic Advisors, Washington, D.C.

Carol Goodloe, Office of Chief Economist, U.S. Department of Agriculture, Washington, D.C.

Mike Guidicipetro, Animal and Plant Health Inspection Service, U.S. Department of Agriculture, San Francisco, California

Michael Hanemann, Department of Agricultural and Resource Economics, University of California, Berkeley

Daniel Hilburn, Department of Agriculture, Oregon State University, Salem

Thomas Hofacker, State and Private Forestry, U.S. Forest Service, Washington, D.C.

Neal Hooker, Center for Food Safety, Texas A&M University, College Station

Linda Horton, International Policy, U.S. Food and Drug Administration, Rockville, Maryland

Stan Kaplan, Bayesian Systems Inc., Rancho Palos Verdes, California

Carol Kramer-LeBlanc, Center for Nutrition Policy and Promotion, U.S. Department of Agriculture, Washington, D.C.

Marsha Kreith, Agricultural Issues Center, University of California, Davis

Mary Lisa Madell, Trade Support Team, International Services, Animal and Plant Health Inspection Service, U.S. Department of Agriculture, Washington, D.C.

Sally McCammon, Animal and Plant Health Inspection Service, U.S. Department of Agriculture, Washington, D.C.

Susan Offutt, Economic Research Service, U.S. Department of Agriculture, Washington, D.C.

Craig Regelbrugge, American Nursery and Landscape Association, Washington, D.C.

Scott Schlarbaum, Department Forestry, Wildlife and Fisheries, University of Tennessee, Knoxville

Steven Shafer, Office of Risk Assessment and Cost-Benefit Analysis, U.S. Department of Agriculture, Washington, D.C.

Katherine R. Smith, Market and Trade Economics, Economic Research Service, U.S. Department of Agriculture, Washington, D.C.

Jan Staman, Nature Management and Fisheries, Ministry of Agriculture, the Hague, the Netherlands

Daniel Sumner, Department of Agriculture and Resource Economics, University of California, Davis

Sue Tolin, Plant Pathology and Weed Science, Virginia Polytechnic Institute and State University, Blacksburg

Trang Vo, Animal and Plant Health Inspection Service, U.S. Department of Agriculture, Riverdale, Maryland

David Wood, Department of Entomology, University of California, Berkeley

Dorothea Zadig, California Department of Food and Agriculture, Sacramento